Communications
in Computer and Information Science   2148

Series Editors

Gang Li⬤, *School of Information Technology, Deakin University, Burwood, VIC, Australia*
Joaquim Filipe⬤, *Polytechnic Institute of Setúbal, Setúbal, Portugal*
Zhiwei Xu, *Chinese Academy of Sciences, Beijing, China*

AF172976

Rationale

The CCIS series is devoted to the publication of proceedings of computer science conferences. Its aim is to efficiently disseminate original research results in informatics in printed and electronic form. While the focus is on publication of peer-reviewed full papers presenting mature work, inclusion of reviewed short papers reporting on work in progress is welcome, too. Besides globally relevant meetings with internationally representative program committees guaranteeing a strict peer-reviewing and paper selection process, conferences run by societies or of high regional or national relevance are also considered for publication.

Topics

The topical scope of CCIS spans the entire spectrum of informatics ranging from foundational topics in the theory of computing to information and communications science and technology and a broad variety of interdisciplinary application fields.

Information for Volume Editors and Authors

Publication in CCIS is free of charge. No royalties are paid, however, we offer registered conference participants temporary free access to the online version of the conference proceedings on SpringerLink (http://link.springer.com) by means of an http referrer from the conference website and/or a number of complimentary printed copies, as specified in the official acceptance email of the event.

CCIS proceedings can be published in time for distribution at conferences or as postproceedings, and delivered in the form of printed books and/or electronically as USBs and/or e-content licenses for accessing proceedings at SpringerLink. Furthermore, CCIS proceedings are included in the CCIS electronic book series hosted in the SpringerLink digital library at http://link.springer.com/bookseries/7899. Conferences publishing in CCIS are allowed to use Online Conference Service (OCS) for managing the whole proceedings lifecycle (from submission and reviewing to preparing for publication) free of charge.

Publication process

The language of publication is exclusively English. Authors publishing in CCIS have to sign the Springer CCIS copyright transfer form, however, they are free to use their material published in CCIS for substantially changed, more elaborate subsequent publications elsewhere. For the preparation of the camera-ready papers/files, authors have to strictly adhere to the Springer CCIS Authors' Instructions and are strongly encouraged to use the CCIS LaTeX style files or templates.

Abstracting/Indexing

CCIS is abstracted/indexed in DBLP, Google Scholar, EI-Compendex, Mathematical Reviews, SCImago, Scopus. CCIS volumes are also submitted for the inclusion in ISI Proceedings.

How to start

To start the evaluation of your proposal for inclusion in the CCIS series, please send an e-mail to ccis@springer.com.

Jitendra Jonnagaddala · Hong-Jie Dai ·
Ching-Tai Chen

Editors

Large Language Models for Automatic Deidentification of Electronic Health Record Notes

International Workshop, IW-DMRN 2024
Kaohsiung, Taiwan, January 15, 2024
Revised Selected Papers

 Springer

Editors
Jitendra Jonnagaddala (ID)
University of New South Wales
Sydney, NSW, Australia

Hong-Jie Dai (ID)
National Kaohsiung University of Science
and Technology
Kaohsiung, Taiwan

Ching-Tai Chen (ID)
Asia University
Taichung, Taiwan

ISSN 1865-0929   ISSN 1865-0937 (electronic)
Communications in Computer and Information Science
ISBN 978-981-97-7965-9   ISBN 978-981-97-7966-6 (eBook)
https://doi.org/10.1007/978-981-97-7966-6

This Springer imprint is published by the registered company Springer Nature Singapore Pte Ltd.
The registered company address is: 152 Beach Road, #21-01/04 Gateway East, Singapore 189721, Singapore

If disposing of this product, please recycle the paper.

Preface

This volume highlights papers carefully selected from the 2024 International Workshop on Deidentification of Electronic Medical Record Notes (2024 IW-DMRN), held as the concluding event to the SREDH/AI Cup 2023 competition on privacy protection and standardization of electronic medical record notes. The competition was hosted by the Intelligent System Lab at the Department of Electrical Engineering, National Kaohsiung University of Science and Technology (NKUST), in collaboration with the Department of Bioinformatics and Medical Engineering at Asia University and the Secure Research Environment for Digital Health (SREDH) consortium in Australia. The IW-DMRN workshop took place on January 15th, 2024, at NKUST, Kaohsiung, Taiwan, R.O.C. The Workshop was organized by the SREDH Consortium, Asia University (Taiwan), NKUST (Taiwan), University of New South Wales (Australia), and AI Cup 2023; and sponsored by Ataraxis AI (USA), CGD Health Pvt Ltd (India), and Ministry of Education (Taiwan).

Electronic medical record (EMR) notes play a pivotal role in medical data analysis, enhancing medication safety, and optimizing medical care efficiency. Moreover, they are instrumental in biomedical research, offering access to real-word datasets of patient information that support retrospective, clinical research, and population health studies. However, to leverage EMR notes effectively for secondary data analysis, it is essential to address patient privacy concerns by removing or de-identifying sensitive data. Additionally, standardizing the representation of time in EMR text notes is crucial, as these notes often portray time in varying formats. Normalizing and unifying temporal information into a standard format facilitates effective comparative analysis.

To tackle these issues, we initiated the 2023 SREDH/AI Cup competition, which consisted of two subtasks: Subtask 1 focused on sensitive health information (SHI) recognition, while Subtask 2 addressed temporal information normalization. Funded by the Ministry of Education, Taiwan, the competition took place between September 18, 2023, and January 8, 2024, using the CodaLab Platform (https://codalab.lisn.ups aclay.fr/competitions/15425). A total of 721 participants from 291 teams engaged in the competition, with 103 teams submitting 218 prediction results for the final test set. Detailed information about the competition is available at https://www.sredhconsort ium.org/sredh-workshops/2024-iw-dmrn . The top 30 teams were invited to submit their manuscript, resulting in the publication of 15 papers in this Springer proceedings volume. All the accepted papers were peer-reviewed by two qualified reviewers chosen from our scientific program committee in a single-blind manner. Furthermore, the top 3 teams were invited to expand their Springer proceedings manuscript for the npj Digital Medicine special collection (https://www.nature.com/collections/fjfiejggae/guest-editors), edited by the volume editor Jitendra Jonnagaddala.

We extended our gratitude to the authors for their insightful contributions and to the scientific program committee members and external reviewers for their valuable comments and feedback during the review process. Special thanks are extended to our keynote speakers, Ming-Iu Lai and Chao-Hsuan Ke, for their enlightening presentations. We acknowledge the Ministry of Education, Taiwan for funding the competition and express our appreciation to our sponsors. We also want to express our gratitude to the SREDH Consortium (www.sredhconsortium.org) Translational Cancer Bioinformatics working group for their support in accessing the OpenDeID corpus v2 Dataset. This project also received funding from the Australian National Health and Medical Research Council (GNT1192469), along with support from Google's research grants and cloud computing resources (GCP19980904). Additionally, we appreciate the assistance from the Research Technology Services at the University of New South Wales Sydney and NVIDIA's academic hardware grant programs.

May 2024   Jitendra Jonnagaddala
                                                                       Hong-Jie Dai
                                                                      Ching-Tai Chen

Organization

General Chair

Jitendra Jonnagaddala   UNSW Sydney, Australia

Organizing, Editorial Committee and Program Committee Chairs

Jitendra Jonnagaddala   UNSW Sydney, Australia/NMC Royal Hospital,
                                             United Arab Emirates
Hong-Jie Dai   NKUST, Taiwan
Ching-Tai Chen   Asia University, Taiwan

Scientific Program Committee

Jan Witowski   Ataraxis AI, USA
Hao-Ping Yang   National Kaohsiung University of Science and
                                             Technology, Taiwan
Hsin-Min Wang   Institute of Information Science, Academia
                                             Sinica, Taiwan
Shalini Gupta   CGD Health, India
Wan-Shu Cheng   Providence University, Taiwan
Omkar Panchal   CGD Health, India
Zheng-long Wu   Soochow University, Taiwan

Keynote Speakers

Ming-Iu Lai   ASUS, Taiwan
Chao-Hsuan Ke   Innolux Corporation, Taiwan

Additional Reviewers

Hsin-Min Wang   Academia Sinica, Taiwan
Wan-Shu Cheng   Providence University, Taiwan
Zheng-Long Wu   Soochow University, Taiwan

Organizers

SREDH Consortium

Asia University, Taiwan

University of New South Wales, Australia

National Kaohsiung University of Science and Technology, Taiwan

AI-Cup 2023

Sponsors

Ataraxis AI, New York, USA

CGD HEALTH Pvt. Ltd., India

Ministry of Education, Taiwan

Contents

Deidentification and Temporal Normalization of the Electronic Health Record Notes Using Large Language Models: The 2023 SREDH/AI-Cup Competition for Deidentification of Sensitive Health Information

Tatheer Hussain Mir[1] , Hao-Ping Yang[1] , Yi-Yun Chou[2] , Yu-Chin Teng[2] ,
Wei-Hsiang Liao[2] , Yu-Chuan Lin[2] , Shalini Gupta[9] , Omkar Panchal[9] ,
Jitendra Jonnagaddala[4,5(✉)] , Ching-Tai Chen[2,3(✉)] , and Hong-Jie Dai[1,6,7,8(✉)]

[1] Intelligent System Laboratory, College of Electrical Engineering and Computer Science,
National Kaohsiung University of Science and Technology, Kaohsiung, Taiwan
{F109154156,hjdai}@nkust.edu.tw

[2] Department of Bioinformatics and Medical Engineering, Asia University, Taichung, Taiwan
ctchen@asia.edu.tw

[3] Center for Precision Health Research, Asia University, Taichung, Taiwan

[4] University of New South Wales, Sydney, Australia
z3339253@unsw.edu.au

[5] NMC Royal Hospital, Khalifa City, Abu Dhabi, United Arab Emirates

[6] National Institute of Cancer Research, National Health Research Institutes, Tainan, Taiwan

[7] School of Post-Baccalaureate Medicine, College of Medicine, Kaohsiung Medical University,
Kaohsiung, Taiwan

[8] Center for Big Data Research, Kaohsiung Medical University, Kaohsiung, Taiwan

[9] CGD Health Pvt. Ltd., Hyderabad, Telangana, India
{shalini,omkar}@cgdhealth.com

Abstract. Electronic Medical Records (EMR) implementation benefits the medical industry with streamlined data analysis, increased patient medication safety, reduced expenses for pathology report storage, and improved medical care efficiency. The EMR text notes hold a patient's clinical record, comprising notes written by the medical staff and further analyzed by the doctor following diagnosis. However, utilizing the EMR text note in its raw form can expose sensitive personal data belonging to patients and medical personnel. Therefore, safeguarding this private information is of utmost importance. Furthermore, the expression of time information in EMR text notes varies across institutions, which can significantly impact the accuracy and reliability of temporal information analysis. Therefore, normalizing temporal information is also a critical issue. The study presents a competition titled Privacy Protection and Standardization of Electronic Medical Record Competition that addresses recognizing Sensitive health information (SHI) recorded in EMR text notes and normalizing temporal information that poses a risk of identity theft. The competition released a corpus containing synthesized SHIs and normalized temporal information. The highest performance for

J. Jonnagaddala et al. (Eds.): IW-DMRN 2024, CCIS 2148, pp. 1–16, 2025.
https://doi.org/10.1007/978-981-97-7966-6_1

the SHI recognition (subtask 1) and temporal information normalization (subtask 2) are micro-/macro-F of 0.949/0.912 and 0.844/0.869, respectively. Overall, the average micro/macro score for subtasks 1 and 2 were 0.666/0.496 and 0.6/0.394.

Keywords: Electronic medical record text notes · sensitive health information · natural language processing · de-identification · temporal information normalization

1 Introduction

Following the enactment and adoption of the Health Insurance Portability and Accountability Act (HIPAA) by the U.S. federal government in 1996, medical establishments worldwide have increasingly embraced electronic health records (EHR) systems. Consequently, EHR has become a fundamental tool for clinical data analysis and has facilitated the digitization of patient records into electronic medical records (EMRs) [1].

To effectively use EMR for secondary data analysis, relevant data involving patient privacy need to be removed (i.e., de-identification). In 2020, the AI CUP (Artificial Intelligence CUP) [2] held a "Medical and Patient Data De-Identification" competition, allowing participating teams to develop models that have made an essential contribution to the de-identification of sensitive health information (SHI) in Chinese conversation data. However, the dataset was compiled based on the Chinese conversation between patients and doctors. Therefore, directly applying the developed model to de-identify EMR text notes written by physicians or other medical staff in English is still a challenge. In addition, the representation of time in EMR text notes is also an important issue. EMR text notes represent time in different ways, such as using different date formats, time formats, time units, and more, which makes it challenging to analyze and compare time information. To conduct practical time information analysis, time needs to be normalized and unified into a standard format or standard unit to facilitate comparison and analysis across data sets and achieve accurate processing of time information in EMR data. Therefore, EMRs should be properly utilized in medical institutions to promote the development of medical research and innovative medical applications [3].

We have hosted a competition, AI CUP 2023-Privacy Protection and Standardization of EMR Competition [4]-focusing on recognizing patients' SHI and normalizing the temporal information [5]. The objective of this challenge is to create a dataset for SHI recognition and temporal information normalization, aiming to advance the field of clinical natural language processing (NLP). In consideration of this, the challenge comprises two subtasks: Subtask 1 involves SHI recognition, while subtask 2 focuses on temporal information normalization. As we know, EMR text notes data must be de-identified to protect personal information. This involves recognizing SHIs and synthesizing their surrogates [6–8]. Similar challenges and datasets [9–11] exist for subtask 1. On the other hand, subtask 2 focuses on extracting and normalizing temporal information, which can aid in accurate diagnosis, treatment planning, and patient monitoring. Furthermore, the duration of untreated symptoms is a critical clinical construct that can be associated with worse intervention outcomes. Estimating the duration of untreated symptoms requires knowing when the symptoms started (symptom onset) and when adequate treatment was initiated. Therefore, a unified standard is needed to convert time formats and units into a consistent format. The dataset compiled for subtask 2 could facilitate the development of NLP systems that can automatically extract relevant temporal information from free text to address complex natural language understanding tasks such as symptom onset extraction, retrospective calculation of the duration of untreated symptoms, and even facilitate the process of temporal surrogate generation during the de-identification process.

2   Methods

The privacy protection and standardization of electronic medical record competition includes two subtasks corresponding to the privacy protection and standardization of EMRs carried out on the same patients' EMR collection based on the extended version of the OpenDeID corpus [12]. Figure 1 shows a segment of a de-identified report sampled from the corpus, wherein the SHIs are highlighted in bold. The text files and the corresponding manually curated annotations, as illustrated in Table 2, form the training and development sets for subtask 1, provided to participants for constructing their systems. A blind test set containing the text files was given only to participants in the testing phase. They submitted predictions and underwent evaluation against manual annotations.

D.O.B: **09/08/2957**
Sex: F
Collected: **14/02/3014 at 11:42**
Location: **3 ARRIETTA CLOSE-POW**
DR **AADLAND ABRAHAM**
CLINICAL:
Lymphoma. Duodenal uptake on PET scan.
Result required for multidisciplinary meeting on **Friday**.

Fig. 1. Example de-identified text report highlighted with SHIs in bold. From the AICUP-2023 Privacy Protection and Standardization of Electronic Medical Record Competition CodaLab website, by ISLab (https://codalab.lisn.upsaclay.fr/competitions/15425#learn_the_details-awards).

Example SHI:

- Type: Date of Birth
- Content: 09/08/2957
- Starting Index in the Report: 8
- End Index in the Report: 18

SHI here corresponds to the Date of Birth information. For example, the text data "09/08/2957" starts at the 8th character and ends at the 18th character in the report. Therefore, the participants have to identify and output SHI details in their predictions based on the specified format and associate them with the respective report identifiers. Similarly, the same follows for the other SHIs. The participants need to output their predictions following the format defined in Table 1 by including the associated report identifier as the first column to prepare their submission file, which can then be submitted to the official CodaLab website for evaluation [27].

Table 1. The manually annotated SHI information for Fig. 1.

SHI content	Starting index in the report	End index in the report	Type
09/08/2957	8	18	DATE
14/02/3014 at 11:42	38	57	TIME
3 CLOSE BACK	69	85	DEPARTMENT
POW	87	89	HOSPITAL
ADLAND ABRAHAM	93	108	DOCTOR
Friday	208	214	DATE

For subtask 2, the participants must output the normalized values for the temporal SHIs, as illustrated in Table 2. The normalized results should comply with the international standard ISO 8601 for representing temporal information. ISO 8601 is a standard that universally accurately represents dates and times. It provides numerical representations for time units, including year, month, day, hour, minute, and second. The standard also permits time intervals and lengths. It standardizes time and date information, decreasing communication challenges and inaccuracies in scientific domains, software development, and worldwide correspondence. Several sectors commonly utilize ISO 8601 and also recommend it for electronic forms.

Table 2. The manually normalized results for the temporal SHIs mentioned in Fig. 1.

Temporal SHI content	Starting index	End index	SHI type	Normalized results
09/08/2957	8	18	DATE	2957-08-09
14/02/3014 at 11:42	38	57	TIME	3014-02-14T11:42
Friday	208	214	DATE	3014-02-18

2.1 Dataset

The data source of the competition is derived from the Health Science Alliance (HSA) biobank of the Lowy Cancer Research Center of the University of New South Wales (UNSW) in Australia [13]. The OpenDeID v2 corpus comprises 3,244 pathological reports out of which 2,100 reports are from the original OpenDeID v1 corpus [14]. OpenDeID corpus is a valuable resource due to its nature for developing and advancing automated patient de-identification systems. The corpus is also the first Australian-based gold-standard corpus explicitly created for this purpose. Furthermore, the corpus is annotated with 38,414 Sensitive Health Information (SHI) entities with high accuracy and deviation scores in inter-annotator agreement. Among the unique characteristics is that it has been manually annotated with surrogate information, maintaining the anonymity of patient information. This dataset's relevance and diversity make it a valuable resource for research endeavors in the healthcare field. We updated the original HSA study SHI corpus annotation guidelines described in the next section to annotate observed SHIs manually. The dataset was partitioned into the training set, validation set, and test set, which consists of 1,734, 560, and 950 reports, respectively.

2.2 Annotation Guidelines

We expanded the original HSA research SHI corpus annotation guide with the subcategories of the DATE category. We applied the normalization procedure for SHIs under the DATE category according to the standard guidelines [15] of ISO 8601. Table 3 presents a detailed breakdown of the 8 major categories and 31 subcategories defined in the annotation guide. Detailed documentation about the annotation guidelines for all defined categories can be found on the ISLab website [15]. Compared to the previously released corpus [12], the corpus compiled for the challenge goes a step further by incorporating detailed temporal annotations and normalization values; we, therefore, provide detailed guidelines for the temporal annotation task—the temporal data, which includes dates, times, durations, and frequencies. Table 5 illustrates example texts and the corresponding normalized values. In the annotation task, a temporal expression is a phrase containing time information. The annotators were asked to annotate any kind of temporal expression within the text. The types of temporal expressions the annotators need to mark include date, time, duration, and frequency. The DATE label is used to annotate dates (absolute and relative dates), the TIME label is used to annotate times and dates, and the DURATION label is used to annotate durations. Finally, the SET label specifically annotates expressions that give quantifiers and intervals (For example, "two times weekly" or "1/day"). A detailed explanation for these aspects is provided below.

The SHI category includes names of patients, doctors, nurses, and others, with three subcategories: Patient, Doctor, and Username. Labeling rules for SHI include highlighting the entire name or initials, not specifying titles, not marking possessive names, using the USERNAME category, and stating the name of professors, registrars, nurses, and others as Doctor. The physician's initials should be indicated.

Location is the SHI category for any mentioned entity location, including hospital name, hospital department, organization name, patient address, city, country, etc. It has multiple subcategories, including sections for address or organization, hospital, department, etc. Healthcare facilities should not be labeled as organizations. Addresses should be labeled, including street, city, state, and zip code. Date is another important category for SHI, as it includes the patient's date of birth and can be used to obtain the patient's age. Dates can be in various formats, such as "/" or "-" as a separator, only day and month, only year, month as number or text, etc. Therefore, annotators need to identify and annotate any format time expression using one of the subcategories. The study focuses on the use of the "i2b2 2012 Temporal Relations Challenge Annotation Guidelines" to determine the representation of "Val." The guidelines outline the importance of age, date, time, duration, and frequency in medical records. Age numbers should be marked, and all ages, not just those over 90, including the patient's family, should be labeled.

Table 3. The SHI categories and sub-categories are defined in the HSA SHI corpus annotation guidelines 5.0.

SHI category	Sub-category	Example		
NAME	PATIENT, DOCTOR, USERNAME	Gudara, Pineau Shaunda, Hinck		
PROFESSION	(none)	Lawyer, teacher		
LOCATION	ROOM, DEPARTMENT, HOSPITAL, ORGANIZATION, STREET, CITY, STATE, COUNTRY, ZIP, OTHER	PERI-OPERATIVE UNIT-POW, MACQUARIE WARD - RHW		
AGE	(none)	23, 98		
DATE	DATE, TIME, DURATION, SET	TEXT	Normalization	Type
		24/12/1987	1987-12-24	DATE
		September 26th	09-26	DATE
CONTACT	PHONE, FAX, EMAIL, URL, IPADDRESS	abc@gmail.com, 194.223.1.1		
IDs	SOCIAL SECURITY NUMBER, MEDICAL RECORD NUMBER, HEALTH PLAN NUMBER, ACCOUNT NUMBER, LICENSE NUMBER, VEHICLE ID, DEVICE ID, BIOMETRIC ID, ID NUMBER	MRN: 9174338 ID NUMBER: 12R1500257		
OTHER	None	Finger Print, Photograph, Company Logo		

Each date subcategory has a "Val" attribute that expresses various forms of normalized values of time information. The purpose of the "Val" field is to store time information in a standard, normalized format so that machines can understand and process it. More specifically, we asked the annotators to adhere to ISO 8601 when formalizing temporal information. Below, we briefly describe the annotation rules.

- DATE "val" representation: ISO 8601 date format is **YYYY-MM-DD. YYYY-MM** is allowed if no exact date info is available. Use only **YYYY** if only year info is available.
- TIME "val" notation: Use the extended time format of ISO 8601 for annotation, specified as **hh:mm:ss.hh**, ranging from 00 to 24. For reduced precision, **ss** and/or **mm** can be omitted, allowing **hh** and **hh:mm** to be used. The combined date and time representation follow the YYYY-MM-DDThh:mm:ss format. Be sure to include

date information in the "val" field if the time expression can be anchored to a specific date.

- DURATION "val" notation: Duration is a time expression that describes a period of time, such as eleven days. The syntax for duration representation is **Pn[YMWD]**. Therefore, "for eleven days" will be represented as "P11D", meaning a period of 11 days. The **n** field does not have to be an integer; "P0.5Y" is valid for half a year. The representations can be combined to represent compound periods, so "P2M3D" represents a period of 2 months and 3 days. As shown in Table 4, the indicator "M" can represent "month" and "minute". To differentiate, we use the "T" designator to represent a time period, such as "P20M" = 20 months, "PT20M" = 20 min.

Table 4. The meaning of the designators used in the DURATION val attributes.

Designator	Meaning	Designator	Meaning
P	Period	T	Time
Y	Year	H	Hour
M	Month	M	Minute
W	Week	S	Second
D	Day		

- SET "val" notation: The frequency of an event can be expressed using ISO 8601 repeat intervals. The ISO repetition interval syntax is R[n][duration], where n represents the number of repetitions and duration represents the DURATION value representation. When n is omitted, this expression represents an unspecified number of repetitions. For example, "once a day for 3 days" is "R3P1D" (repeat a 1-day (P1D) interval 3 times (R3)), and "twice every day" is "RP12H" (repeat every 12 h). In clinical texts we may see frequencies such as p.r.n (taken as needed), which can be represented by "R" (repeated an unspecified number of times for an unspecified duration).

The duration notation is a time expression that describes a period of time, such as eleven days. It can be represented using Pn[YMWD], with the n field not having to be an integer. The frequency of events can be expressed using ISO 8601 repetition intervals, where n represents the number of repetitions and duration represents the duration value representation. When n is omitted, this expression represents an unspecified number of repetitions. Identifiers (ID) are numbers or alphanumeric numbers in text that help identify patients. They can be hospital numbers, record numbers, etc. Annotations for ID categories include:

- Any type of number or alphanumeric should be marked.
- Only IDs with hospital names can be marked as MEDICAL RECORD.
- The block number of the execution program should be marked as IDNUM.
- Lab ID and episode number should be noted as IDNUM.
- When in doubt, label something IDNUM.

- Doctor or nurse ID should be marked as OTHER ID.
- There is no need to label the equipment name.

In the contact category, SHI specifies details that may be useful in contacting the patient, including phone number, email, fax, etc. Annotators should annotate any information in the text record that specifies a patient's contact number. The following rules should be followed when labeling Contact:

- Label all phone numbers, fax numbers, emails, URLs, IP addresses, etc.
- Page numbers should be labeled as telephone numbers.
- The phone number should be usually 4 or 8 digits in the format "02-xxxx-xxxx" and should begin with "029385" because this is the phone code for the hospital to which the record belongs.
- Do not mark numbers starting with #.

Professions should be noted under this category, including occupations like carpenters, teachers, lawyers, etc. The only exception is the medical profession. Positions related to the medical profession should not be labeled. Other information that may help identify the patient but is not included in the above categories should be labeled Other. Examples of this category include fingerprints or company logos. SHI can be images and not just words.

Table 5. Example texts and normalized values for the date subcategories.

Sub-category	Examples (text to be annotated is marked in bold)	Normalized Value
DATE	D.O.B: 16/08/1951	1951-08-16
	(JL/ta 23/12/24)	2024-12-23
	Additional history (... HI-12-111111 3.Apr.82)	2082-04-03
	osteosarcoma identified in the previous resection from	2209
	2209	2221-10
	lesion removed from the left arm in October 2221	2323-07-26 2323-09-16
	... Collected: 26/07/2323 at:... Now having palliative TAH...	
TIME	Collected: 05/09/2094 at 11:42	2094-09-05T11:42
	.. A/Prof E Salisbury at 9:30 am on 18/3/53	2053-03-18T09:30
	Dr N Lambie at 9:16 am on the 12th of September 2713	2713-09-12T09:16
DURATION	... whilst 20weeks pregnant in 2933	P20W
	Lung cancer resected two weeks ago	P2W
SET	... tested twice with positive controls;	R2

2.3 Annotation Process

Four annotators used the multi-document annotation environment (MAE) version 2.1.3 [16] to conduct the annotation process. A configuration file was defined for MAE to enable it to load the textual content of an electronic health record (EHR) text notes with an annotation interface. MAE is utilized due to its ability to handle complex annotation tasks without requiring extensive setup time, and it can load EHR text content and offer an editing interface for annotators to annotate consecutive words as SHI and assign them a relative SHI category. Additionally, the interface allows the annotator to annotate one or more consecutive words as SHIs and assign them to specific SHI categories and the normalized temporal values if the annotated information pertains to the date category. The annotation process is elaborated as follows.

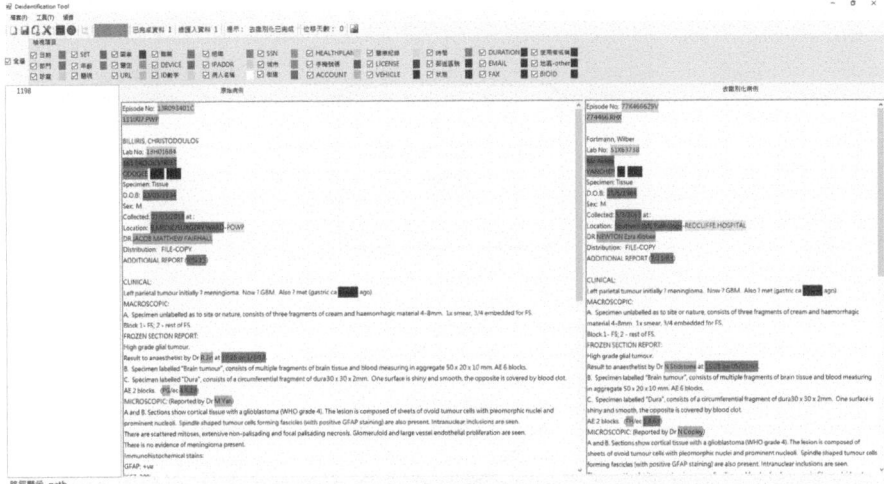

Fig. 2. The ISLab de-identification tool is used for generating surrogates.

Figure 3 represents the dataset creation process of this study. First, four annotators annotated 50 identical EMR text notes each time. Then, they held a meeting after completing the annotation to discuss the problems, concerns, and specific cases requiring discussion during annotation. Finally, they adjusted the annotation guidelines based on the conclusions of the meeting. The annotators followed the procedure for 9 iterations and confirmed the final Kappa inter-annotation agreement [17] to be larger than 0.9. We then divided the remaining 2,794 reports evenly among all annotators for annotation. The compiled corpus was finally processed by the ISLab de-identification tool [18] illustrated in Fig. 2, to generate the surrogates for all annotated SHIs automatically.

2.4   Evaluation Metrics

Both subtasks apply the precision, recall, and micro/macro-F1-measure to evaluate and compare the performance of the participants' predictions on the test set. The evaluation formula definition is described in detail below. For subtask 1, the three attributes of a SHI, including the starting index, ending index, and category, are considered to assess the number of true/false positive/negative cases. On the other hand, for subtask 2, in addition to the aforementioned three attributes, each temporal expression contains the

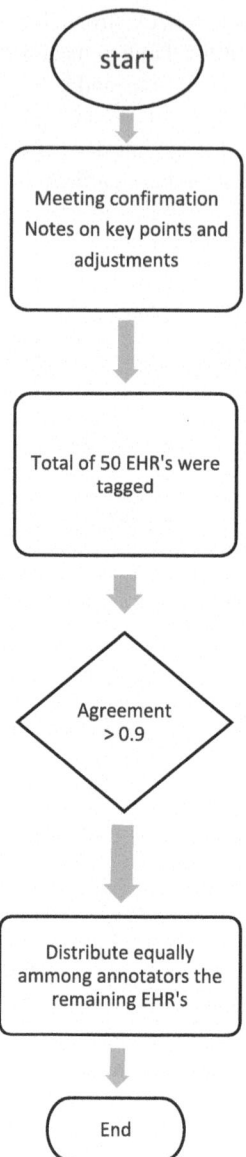

Fig. 3. Data set creation process

normalized result. Based on the above attributes of recognized SHIs, we define a true positive (TP) as accurately extracting all of the three or four attributes annotated by a human annotator for a true SHI; to be considered a TP, the starting and ending indexes of a SHI and its predicted SHI category must match with the manually annotated SHI. The false positive (FP) case occurred when any aforementioned attributes of a predicted SHI cannot match with any manual annotations; 3. False negative (FN): The manual annotated SHIs cannot find an exact matching model prediction result.

$$Precision_{(c)} = \frac{True\ Positives_{(c)}}{\left(True\ Positives_{(c)} + False\ Positives_{(c)}\right)} \quad (1)$$

$$Recall_{(c)} = \frac{True\ Positives_{(c)}}{\left(True\ Positives_{(c)} + False\ Negatives_{(c)}\right)} \quad (2)$$

$$Micro - F_1 - Measure_{(c)} = 2 \cdot \frac{(P \cdot R)}{(P + R)} \quad (3)$$

$$Macro - F_1 - Measure = \frac{2 * \frac{\sum_{i=1}^{n} P_i}{|C|} * \frac{\sum_{i=1}^{n} R_i}{|C|}}{\frac{\sum_{i=1}^{n} P_i + \sum_{i=1}^{n} R_i}{n}} \quad (4)$$

where, C is the category, c is the subcategory, and n is the total number of all occurring subcategories. Participants' predictions for both subtasks were assessed and ranked using macro-F1 measures. Final rankings were determined by averaging individual ranks from both tasks for each participant and sorting the teams based on the averaged values. Take the team "zhaorui" shown in Fig. 4 as an example. The participant secured a fifth rank in the first task and a first rank in the second task. With an averaged rank of three, as reflected in the "<Rank>" column, they attained a leading position on the leaderboard.

Privacy Protection and Medical Data Standardization Competition Reference Ranking

#	User	Entries	Date of Last Entry	<Rank> ▲	子任務 1：病患隱私資訊擷取 ▲	子任務 2：時間資訊正規化 ▲	Detailed Results
1	zhaorui	1	12/02/23	3.00 (1)	0.8660 (5)	0.8685 (1)	View
2	AUSTIN2526	3	12/02/23	4.00 (2)	0.8815 (1)	0.7910 (7)	View
3	henzee	6	12/03/23	4.50 (3)	0.8425 (7)	0.8411 (2)	View
4	Chen128	1	12/02/23	5.00 (4)	0.8789 (2)	0.7770 (8)	View
5	paveen	5	12/03/23	7.50 (5)	0.8094 (9)	0.7959 (6)	View
6	YongZhenHuang	3	12/02/23	8.00 (6)	0.8780 (3)	0.7384 (13)	View
7	Pinyi	6	12/02/23	9.00 (7)	0.7782 (15)	0.8221 (3)	View
8	CHIAYICHAO	3	12/03/23	10.00 (8)	0.7773 (16)	0.8065 (4)	View
8	Xiuyu223	15	12/03/23	10.00 (8)	0.8647 (6)	0.7259 (14)	View
9	Vik	9	12/03/23	11.00 (9)	0.7934 (11)	0.7556 (11)	View
10	seven.tychi	8	12/03/23	14.00 (10)	0.7747 (18)	0.7625 (10)	View
10	allan1123	2	12/03/23	14.00 (10)	0.7875 (13)	0.7113 (15)	View
10	C110133204	2	12/02/23	14.00 (10)	0.8663 (4)	0.6257 (24)	View

Fig. 4. The test set leaderboard on the Codalab website for the top-10 teams.

3 Results

3.1 SHI Category Statistics

The entire 3,244 reports were divided into training, validation, and test sets, which contain 1,734, 560 and 950 reports, respectively. Table 6 shows a summary of compiled corpus in terms of the number of SHI annotations in the training, validation, and test sets.

Table 6. SHI category statistics.

	Training set	Validation set	Test set	Total
PATIENT	1819	545	716	1261
DOCTOR	6541	2191	3327	12059
ROOM	1	0	0	1
DEPARTMENT	1025	366	419	1810
HOSPITAL	1801	593	1198	3592
ORGANIZATION	135	24	74	233
STREET	899	321	344	1564
CITY	939	337	373	1649
STATE	875	310	332	1517
COUNTRY	3	2	0	5
ZIP	913	325	353	1591
LOCATION-OTHER	6	4	6	16
AGE	127	57	51	235
DATE	4753	1622	2459	8834
TIME	1119	420	470	2009
DURATION	23	6	12	41
SET	14	0	5	19
PHONE	9	2	1	12
MEDICALRECORD	1824	563	747	3134
IDNUM	3677	1177	2120	6974

3.2 Overview of the Competition Results

The competition attracted a substantial number of participants, including both teams and individuals, contributing to the overall richness of the dataset and reflecting a robust engagement with the tasks presented. A total of 721 participants joined the competition, constituting 291 teams. Teams played a crucial role in the competition. During the development phase, only the team leader could submit up to three runs per day to evaluate their model performance using the leaderboard. In the final testing phase, each team can submit a maximum of six runs over two days. Out of the participating teams, 103 teams submitted their prediction results for the test set, resulting in a total number of 218 submissions during the final testing phase, reflecting the dynamic and iterative nature of the competition.

The highest micro-/macro-F-scores achieved by participants for subtasks 1 and 2 were 0.949/0.912 and 0.844/0.869, respectively. Examining the overall performance, the average micro-/macro-score for subtasks 1 and 2 stood at 0.666/0.496 and 0.6/0.394, indicating a substantial competency level among participants and various approaches and solutions. These findings contributed to a nuanced understanding of participant performance and the effectiveness of the competition in fostering excellence and innovation within the specified tasks. This provides quantification of the extent of participant involvement and offers valuable insights into the distribution of scores, the diversity of approaches, and the overall success achieved in meeting the competition's objectives.

4 Discussion

One NLP trend noticed in the competition is that 77.2% of participants used large language models (LLMs) in their systems based on the online questionnaires filled out by 58 participants. Upon further breaking down the use of LLM architectures, the most prevalent was decoder-only transformer architecture, comprising 68.9% of total instances. The encoder-decoder architecture was the next most common, accounting for 26.7%, while the encoder-only architecture was utilized only by 4.4% of the instances. This distribution provides a clear overview of the relative prevalence of different architectures in the given context. Furthermore, the EleutherAI/Pythia models emerged as a popular choice. Here are some highlights of the top-performing team we would like to discuss.

TEAM_3970 (CodaLab ID: zhaorui), ranking 1st in the competition, performed a comparative study that assessed the efficacy of two approaches: fine-tuning a large language model based on the Chat generative pre-trained transformer (ChatGPT) and a rule-based approach. The ChatGPT-based approach achieved high macro-F-scores for SHI recognition and temporal information normalization, while the rule-based approach demonstrated lower latency and power consumption. TEAM_4761, with CodaLab ID C110133204 and ranked 2nd, utilized large language models and in context-learning algorithms to ensure correct model inference and to utilize invalid label features. The research utilized large language models for deidentification and temporal normalization, adopting a sliding window and masking strategies. To overcome the overfitting in the embedding layer, they introduced NEFtune technology in their study for improved learning conditions. The study successfully identified and de-identified sensitive information using these powerful models. Team_4593, Codalab ID: Paveen, which ranked 3rd, explored using discriminative and generative models to improve privacy and standardize medical data. It found that generative models outperformed traditional discriminative models in PHI extraction tasks. The research also introduced diverse data forms, small sample augmentation, dual-model fusion, and a voting mechanism. Future research aims to explore finer-grained loss functions, advanced model fusion techniques, and PEFT techniques. Team_TMUNLP developed a textual data model for clinical entity recognition using a stringent preprocessing protocol. They refined the T5-Efficient-BASE-DL2 model for computational efficiency and structured the dataset into two training sets: one validation set and a unique test set. The AutoTokenizer used AutoModel-ForSeq2SeqLM classes from the Hugging Face library for consistency. The model's performance was assessed without gold standard annotations, highlighting the value of rule-based methods in clinical entity recognition. TEAM_3868, with CodaLab ID: vik, presented a dual-model fusion with a voting mechanism, diverse data forms, and data augmentation techniques, demonstrating the superiority of generative models over traditional discriminative models in PHI extraction and TIN tasks, paving the way for advanced AI applications in healthcare.

Furthermore, many other teams used different approaches while solving this problem; one team developed a textual data model for clinical entity recognition using a stringent preprocessing protocol and refined the T5-Efficient-BASE-DL2 model for computational efficiency. Another team based their model architecture on the Transformer framework, employing rule-based postprocessing to address low recall rates, particularly for unpopular labels. They applied rule-based processing to seven labels in the competition and expanded it to include other unpopular and popular labels, resulting in improved scores. Another team developed an algorithm using Inverse Correlational Learning (ICL) and the Sliding Windows method to improve the correlation between sequences and their labels. Another team used a rule-based NER system to categorize annotation files using HashMap and HashSet. Another team developed a hybrid approach to recognize and standardize SHIs, using the Pythia language model and a regular expression program to enhance recognition accuracy. Another team used model architecture based on the transformer framework, incorporating GPTNeoXAttention, GPTNeoXRotaryEmbedding, parallel computation of Attention and Feed-Forward, and all Dense Layers. They used the "AdamW" optimization method to enhance the Adam algorithm and employed rule-based postprocessing to address low recall rates, particularly for unpopular labels. Another team developed an algorithm using Inverse Correlational Learning (ICL) and the Sliding Windows method to improve the correlation between sequences and their labels. The COT-few method was integrated into the Prompting approach to address the small number of labels in the DURATION and SET categories. Another team used advanced pre-trained models and evaluated discriminative and generative approaches for SHI extraction and time information normalization in medical records. The Discharge Summary BioBERT hybrid de-identification model demonstrated the ability to maintain a delicate balance between preserving patient privacy and ensuring accurate deidentification.

5 Conclusion

The goal of the AICUP-2023 Privacy Protection and Standardization of Electronic Medical Record Competition is to promote the development of applying state-of-the-art NLP technologies like LLMs on two subtasks: the de-identification task and the temporal information normalization task. An OpenDeID corpus V2 containing 3,244 pathology reports in the Health Sciences Alliance Biobank of the University of New South Wales and Lowe Cancer Research Center in Australia was released as a data set for the competition. We summarized the annotation process, the statistics of the released dataset, and the results achieved by all participating teams for the two subtasks. We observed a significant trend in the field of NLP, where an increasing number of researchers and competition participants were focusing on utilizing LLMs. A substantial 77.2% of competition participants prominently incorporated LLMs into their approaches, particularly the decoder-only architecture, giving a sense of recognizing the potential of pre-trained LLM models in health care infrastructure. Overall, the conclusions highlight a strategic and informed engagement in the AI CUP, leveraging advanced AI techniques for meaningful healthcare data management and privacy advancements.

Acknowledgments. This study was funded by Ministry of Education and by the National Science and Technology Council [NSTC 112-2221-E-992 -056 -MY3]. Our appreciation also goes to the University of New South Wales, Lowy Cancer Research Centre, and Health Science Alliance Biobank for granting access to the medical records utilized in this study. Additionally, we recognize the IW-DMRN workshop (https://www.sredhconsortium.org/sredh-competitions/sredhai-cup-2023/2024-iw-dmrn) and the SREDH Consortium (https://www.sredhconsortium.org/) Translational Cancer Bioinformatics working group in accessing the OpenDeID corpus v2 Dataset and for their valuable contributions to our research endeavors JJ was funded by the Australian National Health and Medical Research Council (GNT1192469) and supported by 2022 Google's research grants and cloud computing resources (GCP19980904), as well as the Research Technology Services at University of New South Wales Sydney and NVIDIA's academic hardware grant programs.

Disclosure of Interest. The authors have no competing interests to declare that are relevant to the content of this article.

References

1. Zirikly, A., et al.: Information extraction framework for disability determination using a mental functioning use-case. JMIR Med. Inform. **10**(3), e32245
2. ALDEA. MOE AI competition and labeled data acquisition project. (ALDEA) (2020). https://en.aicup.tw/ai-cup-2020
3. Amber, S.: MAE and MAI: lightweight annotation and adjudication tools. In: 2011 Association for Computational Linguistics, pp. 129–133, June 2011
4. Entity enhanced BERT pre-training for Chinese NER. In: Proceedings of the 2020 Conference on Empirical Methods in Natural Language Processing, Association for Computational Linguistics (2020)
5. Chen, A., Jonnagaddala, J., Nekkantti, C., Liaw, S.T.: Generation of surrogates for de-identification of electronic health records. In: MEDINFO 2019: Health and Wellbeing e-Networks for All, pp. 70–73. IOS Press (2019)
6. Congress, 1. Health Insurance Portability and Accountability Act of 1996. ASPE - office of the assistant secretary for planning and evaluation, 20 August 1996
7. Consortium, S.: 2024 International workshop on deidentification of electronic medical record notes (IW-DMRN). 2024. https://www.sredhconsortium.org/sredh-competitions/sredhai-cup-2023/2024-iw-dmrn
8. Devlin, J., Chang, M.-W., Lee, K., Toutanova, K.: BERT: pre-training of deep bidirectional transformers for language understanding. CoRR (2018)
9. Moharasan, G., Ho, T.B.: Extraction of temporal information from clinical narratives. J. Healthc. Inform. Res. **3**(2), 220–244 (2019)
10. ISLab. AICUP 2023 Privacy Protection and Standardization of Electronic Medical Record Competition, September 2023. Retrieved from Codalab. https://codalab.lisn.upsaclay.fr/competitions/15425#learn_the_details
11. ISLab. HSA Study PHI Corpus - Annotation guidelines (2023). https://nkustislab.github.io/docs/project/aicup/guideline.pdf
12. Islab. De-identification tools (n.d.). https://nkustislab.github.io/#projectModal2
13. Jonnagaddala, J., Chen, A., Batongbacal, S., Nekkantti, C.: The OpenDeID corpus for patient de-identification. Sci. Rep. **11**(1), 19973 (2021)
14. Kajiyama, K., Horiguchi, H., Okumura, T., Morita, M., Kano, Y.: De-identifying free text of Japanese electronic health records. J. Biomed. Semant. **11**(1), 1–12 (2020)

15. Lafferty, A.M.: Conditional random fields: probabilistic models for segmenting and labeling sequence data (2001). https://repository.upenn.edu/entities/publication/c9aea099-b5c8-4fdd-901c-15b6f889e4a7

16. Li, L., Jin, L., Jiang, Z., Song, D., Huang, A.D.: Biomedical named entity recognition based on extended recurrent neural networks. In: 2015 IEEE International Conference on Bioinformatics and Biomedicine (BIBM), Washington, DC, USA (2015). https://ieeexplore.i

17. Morwal, S., Jahan, N., Chopra, D.: Named entity recognition using Hidden Markov Model (HMM). Int. J. Nat. Lang. Comput. (IJNLC) 1(4), 15–23 (2012). https://papers.ssrn.com/sol3/papers

18. Quinn, C.M., et al.: Moving with the times: The Health Science Alliance (HSA) Biobank, pathway to sustainability. Biomarker Insights 16 (2021)

19. Shaalan, K., Khaled: Rule-based approach in Arabic Natural Language Processing. Int. J. Inf. Commun. Technol. 3, 11 (2010). https://m.marefa.org/w/images/1

20. Silvestri, S., Esposito, A., Gargiulo, F., Sicuranza, M., Ciampi, M., De Pietro, G.: A big data architecture for the extraction and analysis of EHR data. https://ieeexplore.ie

21. Stubbs, A., Kotfila, C., Uzuner, O.: Automated systems for the de-identification of longitudinal clinical narratives: overview of 2014 i2b2/UTHealth shared task Track 1. J. Biomed. Inform. 58, S11–S19 (2015)

22. Uzuner, O., Luo, Y., Szolovits, P.: Evaluating the state-of-the-art in automatic de-identification. J. Am. Med. Inform. Assoc. 14(5), 550–563 (2007)

23. Vaswani, A., et al.: Attention is all you need. In: Advances in Neural Information Processing Systems. Long Beach, CA, USA: 31st Conference on Neural Information Processing Systems (NIPS 2017) (2017)

24. Viani, N., et al.: Temporal information extraction from mental health records to identify the duration of untreated psychosis. J. Biomed. Semant. 10(2) (2020)

25. Viera, A.J., Garrett, J.M.: Understanding interobserver agreement: the kappa statistic. Fam. Med. 37(5), 360–363 (2005)

26. Wang, C.K., et al.: Principle-based approach for the de-identification of code-mixed electronic health records. IEEE (2022)

27. Liu, J., et al.: OpenDeID pipeline for unstructured electronic health record text notes based on rules and transformers: deidentification algorithm development and validation study. J. Med. Internet Res. 25, e48145 (2023). https://doi.org/10.2196/48145

28. Consortium, S.: Secure Research Environment for Digital Health (SREDH) Consortium. https://www.sredhconsortium.org/

29. Alla, N.L.V., et al.: Cohort selection for construction of a clinical natural language processing corpus. Comput. Methods Programs Biomed. Update 1, 100024 (2021)

Enhancing Automated De-identification of Pathology Text Notes Using Pre-trained Language Models

Yong-Zhen Huang[1,2] , Tzu-Cheng Peng[1] , Heng-Yu Lin[1] , Eugene Sy[1] ,
and Yung-Chun Chang[1,3(✉)]

[1] Graduate Institute of Data Science, College of Management, Taipei Medical University,
Taipei, Taiwan
{m946111005,m946111003,m946111008,m946111012,changyc}@tmu.edu.tw
[2] Department of Nursing, National Taiwan University Cancer Center, Taipei, Taiwan
[3] Clinical Big Data Research Center, Taipei Medical University Hospital, Taipei, Taiwan

Abstract. With the recent advancements in artificial intelligence (AI), particularly the surge in large language models (LLMs) led by industry giants such as OpenAI, Microsoft, and Google, a new era in clinical medicine and digital health has been ushered in. These advancements, transformative as they are, introduce significant privacy concerns, particularly in the healthcare sector, which heavily relies on Electronic Health Records notes (EHRs) text notes. Our research directly addresses these privacy challenges by enhancing the T5-Efficient-BASE-DL2 model for the de-identification of patient data. We fine-tuned the T5 model with specific hyperparameter adjustments and implemented exact match evaluations to significantly reduce false positives and improve the accuracy of entity recognition, particularly for sensitive information processing. This was achieved using regular expressions and the Pycountry library. Our enhancements resulted in a notable improvement in the model's performance, with a 14% increase in the macro-average F1-score for Task 1, rising from 0.7787 to 0.9187, and a 30% improvement for Task 2, from 0.4871 to 0.79. These advancements not only enhance the security and efficiency of healthcare information management but also uphold the integrity of individual privacy and facilitate the safe sharing of data for research purposes without compromising confidentiality. Our research lays the groundwork for future developments in improving rule-based methodologies and expanding datasets, aiming to further integrate machine learning with linguistic rules for more effective data privacy solutions in healthcare.

Keywords: Large Language Models · Electronic Health Records notes · Sensitive Health Information

1 Introduction

The swift growth of artificial intelligence (AI), particularly in the field of large language models (LLMs), marks a notable shift in technology, with industry leaders such as OpenAI, Microsoft, and Google driving this change by incorporating LLMs like

J. Jonnagaddala et al. (Eds.): IW-DMRN 2024, CCIS 2148, pp. 17–29, 2025.
https://doi.org/10.1007/978-981-97-7966-6_2

ChatGPT into a wide range of products [1–3]. These advancements highlight the considerable capabilities of LLMs across various sectors, especially in clinical medicine, where they're seen as a key advancement in digital health [4, 5]. Research in this domain is growing, fueled by the potential of LLMs to transform healthcare services and medical research. Nonetheless, this technological advancement raises serious privacy concerns. Users and developers may not fully grasp the risks of privacy breaches when using LLMs. These advanced models, trained on extensive datasets that may contain sensitive personal data, could unintentionally expose private information due to their recall and interactive features [6]. This problem is not confined to individual users but also pertains to the application of AI in sensitive areas such as healthcare. The integration of Electronic Health Records notes (EHRs) in medical and biomedical institutions worldwide amplifies this challenge. EHRs, laden with confidential patient details, pose a considerable risk. The scattered data across different EHR systems, if patient information is not sufficiently protected, it could potentially result in patient reidentification, which would compromise their privacy [7].

In response to these critical challenges, the Ministry of Education in Taiwan has sponsored the Artificial Intelligence CUP 2023 - Privacy Protection and Medical Data Standardization Challenge [28]. This initiative is part of the nationwide project "Ministry of Education Artificial Intelligence Competition and Annotation Data Collection Project." The competition aims to attract researchers worldwide to develop solutions for the automatic de-identification and standardization of unstructured text notes, particularly in the context of healthcare data. Participants in this challenge are tasked with evaluating their AI models on a large, multicenter corpus from Australia [8–14].

This study introduces new methods in the field of patient data privacy processing. By integrating the advanced T5-Efficient-BASE-DL2 model with optimized hyperparameter settings and adopting a novel exact-match evaluation strategy, we have markedly improved the model's ability to identify sensitive patient data. We tackled the challenges of certain entity types by introducing a set of rule-based postprocessing techniques. Our specialized approach for handling phone numbers, geographic identifiers, and time expressions not only reduces incorrect classifications but also transforms data into a uniform and practical format. These improvements demonstrate our dedication to technical expertise and establish a new standard for precision and effectiveness in the de-identification of patient data, offering a significant enhancement to privacy protection in healthcare data management.

2 Material and Methods

2.1 Dataset

In this study, we utilize the OpenDeID v2 corpus, a notable dataset in the field of patient privacy and data security. Recognized as the first Australian-based standard resource focused on patient de-identification, this corpus serves as a foundational element for creating and enhancing automated systems aimed at removing personal details from patient records. These systems accommodate both rule-based and machine-learning approaches. The OpenDeID v2 corpus comprises a detailed set of 3,244 pathology reports from 1,833 cancer patients, with an average of about 717 tokens per report (initially including 2,100 reports f and subsequently supplemented with 1,144 additional reports). A key feature of this dataset is its rigorous annotation process, through which 38,414 Sensitive Health Information (SHI) entities have been identified. The reliability of this process is evident from the high inter-annotator agreement and deviation scores, recorded at 0.9464 and 0.9503, respectively. The corpus has been carefully annotated with substitute information to ensure the thorough de-identification of patient data. Consequently, the OpenDeID v2 corpus serves as an essential resource for enhancing de-identification technologies and stands as a strong example of maintaining strict privacy protocols in the management of patient data [13, 14].

2.2   Data Preprocessing

To optimize this corpus for deep learning applications, we applied a stringent preprocessing protocol to the textual data. The process began with the development of an algorithm designed to determine the starting index of each line of text, a critical step for ensuring accurate text segmentation. This setup was crucial in facilitating effective information extraction, laying the groundwork for the subsequent application of advanced learning algorithms. The parsing phase involved formulating prompts for each document to direct the extraction of information compliant with the Health Insurance Portability and Accountability Act (HIPAA) [15]. These prompts began with "Please extract HIPAA-related information from the given text:", followed by the document identifier, the starting index, and the specified text content. In the completion, we identified the type of information and relevant patient privacy data, marking the absence of such data as "PHI: NULL". For temporal data, normalization to a standard format was applied, denoted as "(Normalized: [time normalization content])". These preprocessing steps were crucial for transforming the data to meet the specifications of the Hugging Face's Dataset library, ensuring compatibility with advanced learning algorithms [16]. The processed data was saved in this format, laying the groundwork for a robust framework for subsequent model training, validation, and testing phases. This approach guarantees that the data is structured and optimized for integration with complex models, providing a solid foundation for their evaluation and refinement.

After the model's data predictions, a postprocessing step was enacted, involving the storage of outcomes in a JSONL structure. A custom algorithm was designed to extract and format key information from the JSONL documents. This algorithm was tasked with the retrieval of HIPAA information types and patient confidentiality data from the completions, and the file identifiers and initiation indices from the prompts. It accurately calculated the definitive initiation and cessation indices of the confidentiality data, extrapolating from the extracted content and its intrinsic context and position within the prompts. For temporal data classifications, such as DATE, TIME, SET, and DURATION, an additional extraction of temporal normalization details was executed. The culmination of this process was the synthesis of the extracted content, in accordance with the prescribed format and submission protocols of the task at hand.

2.3   Implementation the T5-Efficient-BASE-DL2 Model for Textural Generation

Within the scope of this study, we engaged the T5-Efficient-BASE-DL2 model—a specialized iteration within the Text-to-Text Transfer Transformer (T5) series, refined for heightened computational efficiency and optimized memory usage. This model, hosted on the Hugging Face platform and developed by Google, is predicated upon the original T5 architecture, and employs an Encoder-Decoder framework adept at executing a multitude of Natural Language Processing (NLP) tasks, with a particular emphasis on text-to-text transformations. The model redefines all NLP challenges as text generation problems, thus standardizing the solution approach. The Efficient-BASE-DL2 variant introduces modifications to the base model, which are specifically designed to enhance performance in environments with limited computational resources. These modifications include reductions in model size and refinements in the computational processes, which collectively contribute to a robust execution with a lower computational footprint [17, 18].

For our research purposes, the model was tasked with processing data formatted as textual input-output pairs (refer to Figure 1 for an illustrative example). Following a thorough preprocessing phase, the data was serialized into JSONL format, ensuring uniformity and standardization across the dataset. The selection of hyperparameters was the result of extensive experimentation, with a batch size of 4, a learning rate of $2e-5$, and a weight decay parameter set to 0.05. The training regimen spanned 10 epochs and included the use of a linear learning rate scheduler to optimize the convergence process. The output generation was constrained to a maximum length of 50 tokens per instance, with evaluations and model checkpoints implemented at the conclusion of each training epoch. For performance evaluation, we utilized the Rouge metric to quantitatively assess the similarity between the generated text and the reference corpus [19], alongside an Exact Match metric to measure the precision of the model's outputs against the expected results.

2.4   Model Training and Validation Strategy

In our study, we adopted a comprehensive approach to data preparation and model training, aimed at developing a robust model with strong generalization capabilities. The dataset was methodically structured into two separate training sets, one validation set, and a unique test set, the latter of which did not contain standard answers and was used exclusively for result submission. During the initial validation phase, the model underwent training using the two distinct training sets. This phase was critical in laying the groundwork for the model's learning process. Following the training, the model was evaluated using the validation set. This step was essential to assess the initial generalization and predictive capabilities of the model, providing insights into its performance on unseen data. For the testing phase, we adopted a strategy of merging the two training sets with the validation set, creating a consolidated dataset for final model tuning. This dataset was then divided using the train_test_split function from the Scikit Learn library [20]. We allocated 90% of the data for training and the remaining 10% for validation.

This allocation was chosen to maximize the model's exposure to a diverse array of data while retaining a significant subset for validation. Such an approach was instrumental in fine-tuning the model's hyperparameters, ensuring that the model was comprehensively trained and capable of generalizing well to new data.

Throughout the process, we employed the AutoTokenizer and AutoModel-ForSeq2SeqLM classes from the Hugging Face library [21]. These tools were integral in efficiently tokenizing the text data and constructing the sequence-to-sequence learning model. Additionally, a custom function was developed to ensure uniform tokenization across the dataset, promoting consistency in data handling. The Seq2SeqTrainer class was utilized to effectively manage the training process, overseeing batch processing, metric evaluation, and model saving [21]. After multiple training cycles and hyperparameter tuning, the model exhibited commendable performance benchmarks. In the prediction phase, the fine-tuned T5-Efficient-BASE-DL2 model was deployed for generating text outputs. We addressed the issue of content truncation, identified during training, by adjusting the maximum generation length to 100 tokens in the prediction phase, thereby enhancing the accuracy and completeness of the generated outputs. This approach to model development and fine-tuning resulted in a model that not only demonstrated strong learning capabilities during the training phases but also excelled in managing longer text sequences during prediction. The model achieved high levels of predictive accuracy and practical relevance, highlighting the efficacy of our comprehensive training and validation strategy.

2.5 Rule-Based Data Extraction and Temporal Normalization

To extract structured data from unstructured text, we implemented a rule-based strategy focusing on identifying various types of identifiable information, as illustrated in Fig. 1. For phone number extraction, our algorithm applied a regex pattern recognizing the local formatting convention: eight digits split into two groups of four, divided by a space. For location data, another regex pattern was used to pinpoint postal and geo-graphic identifiers, for instance, "PO BOX" followed by digits or specific place names. City name extraction harnessed the comprehensive list from the Pycountry library [22], with a supplementary exclusion list to prevent common misidentifications such as mistaking "South Australia" or "Congo Red" for geographical locations. This combination significantly lowered the incidence of false positives. Through these customized rule-based algorithms, we achieved accurate extraction of phone numbers, geographical locations, and city names, essential for the integrity of our data analysis.

For temporal normalization, we devised a conversion algorithm that translated diverse temporal descriptions into their numerical counterparts. This ranged from simple figures to more complex intervals like "two months", "three years", or "four weeks". A broad array of regex rules was established to detect and convert these expressions, enabling effective parsing and reformatting into a structured format.

Example: For a text mentioning "The patient will return in two weeks for a follow-up", our system identifies "two weeks" and normalizes it to "P2W" in accordance with ISO 8601 standards [23], ensuring consistent temporal representation throughout the dataset. Temporal expressions were systematically converted to standardized formats, such as 'P2M' for two months and 'P3Y' for three years. This rigorous normalization, including the conversion of terms like 'once' to 'R1' and 'twice' to 'R2', ensures temporal data is coherently formatted, facilitating seamless integration with sophisticated analytical models.

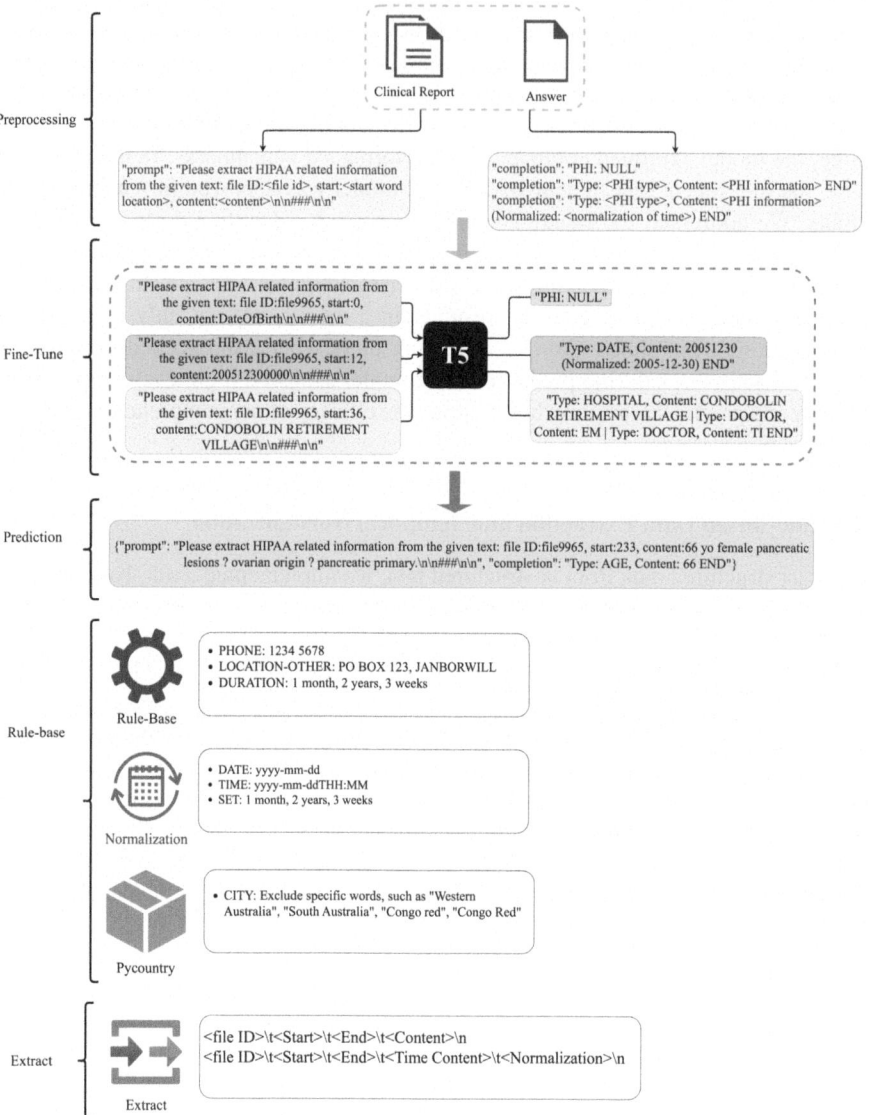

Fig. 1. Example of system architecture and processes involved in the algorithm.

3 Results

In our study, we assessed the model's performance in scenarios lacking gold standard annotations, utilizing the validation dataset to gauge predictive accuracy. Initially, our non-rule-based model faced difficulties in identifying certain entity types, notably COUNTRY and LOCATION-OTHER, due to a scarcity of representative samples. Additionally, the generation of DURATION entities exhibited significant inconsistencies in time normalization. To address these challenges, we implemented rule-based modifications and advanced normalization techniques, achieving substantial improvements in the model's performance metrics. For Task 1, the macro-average F1-score improved by 14%, rising from 0.7787 to 0.9187. Task 2 saw a more pronounced enhancement, with a 30% increase in the F1-score, from 0.4871 to 0.79.

Tables 1 and 2 provide a comprehensive comparison of performance metrics across validation and test datasets for both tasks. The precision in identifying temporal entities like DATE and TIME was commendable, although the macro-average F1 scores for Task 1 (0.8780) and Task 2 (0.7384) indicate potential for improved recall. The high precision but lower recall for DURATION and SET entities underscore the rule-based system's efficacy in accurate entity recognition. However, a slight dip in the macro-average F1-score for Task 1 when transitioning from validation to test data suggested a minor inconsistency in performance across entity types. For Task 2, the disparity between validation and test data was more pronounced, underscoring the need for enhancements to bolster recall and overall F1 scores.

Table 1. Comparative Analysis of Non-Rule-Based and Rule-Based Prediction Models in Clinical Entity Recognition Using Validation Data

Non-Rule-Based Prediction Model		Rule-Based Prediction Model	
Task 1			
Coding Type	*Precision/Recall/F1-score*		*Support*
HOSPITAL	0.9642/0.9545/0.9593	0.9642/0.9545/0.9593	593
PATIENT	0.9737/0.8844/0.9269	0.9737/0.8844/0.9269	545
IDNUM	0.7955/0.8624/0.8276	0.7961/0.8624/0.8279	1177
MEDICALRECORD	0.9778/0.8615/0.9160	0.9778/0.8615/0.9160	563
TIME	0.9226/0.7129/0.8043	0.9249/0.7368/0.8202	418
DOCTOR	0.9192/0.9612/0.9398	0.9197/0.9612/0.9400	2191
DATE	0.9843/0.8970/0.9387	0.9850/0.8964/0.9386	1612
AGE	0.9298/0.9298/0.9298	0.9444/0.8947/0.9189	57
ORGANIZATION	0.6667/0.6667/0.6667	0.6667/0.6667/0.6667	24
CITY	0.9755/0.9436/0.9593	0.9845/0.9436/0.9636	337

(*continued*)

Table 1. (*continued*)

Non-Rule-Based Prediction Model		Rule-Based Prediction Model	
STATE	0.9902/0.9806/0.9854	0.9902/0.9806/0.9854	310
ZIP	0.9878/0.9969/0.9923	0.9908/0.9969/0.9939	325
DEPARTMENT	0.9738/0.9126/0.9422	0.9738/0.9126/0.9422	366
STREET	0.9772/0.8006/0.8801	0.8333/0.8333/0.8333	6
DURATION	0.5000/0.1667/0.2500	0.9772/0.8006/0.8801	321
PHONE	1.0000/1.0000/1.0000	1.0000/1.0000/1.0000	2
COUNTRY	0.0000/0.0000/0.0000	1.0000/1.0000/1.0000	2
LOCATION-OTHER	0.0000/0.0000/0.0000	1.0000/1.0000/1.0000	4
Micro-avg. F	0.9322/0.9044/0.9181	0.9332/0.9064/0.9196	8853
Macro-avg. F	0.8077/0.7517/0.7787	0.9390/0.8992/0.9187	8853
Task 2			
Coding Type	*Precision/Recall/F1-score*		*Support*
TIME	0.8818/0.6244/0.7311	0.8818/0.6244/0.7311	418
DATE	0.7673/0.6873/0.7251	0.7672/0.6867/0.7247	1612
DURATION	0.0000/0.0000/0.0000	1.0000/0.8333/0.9091	6
Micro-avg. F	0.7863/0.6724/0.7249	0.7873/0.6744/0.7265	2036
Macro-avg. F	0.5497/0.4372/0.4871	0.8830/0.7148/0.7900	2036

Table 2. Performance Evaluation of Rule-Based Prediction Models on Test Data in Clinical Entity Recognition

Rule-Based Prediction Model		
Task 1		
Coding Type	*Precision/Recall/F1-score*	*Support*
IDNUM	0.8025/0.8759/0.8376	2120
MEDICALRECORD	0.7565/0.8983/0.8213	747
HOSPITAL	0.8913/0.8623/0.8765	1198
PATIENT	0.9684/0.9413/0.9547	716
DATE	0.9712/0.9199/0.9449	2459
DOCTOR	0.9805/0.9375/0.9585	3327
ORGANIZATION	0.7703/0.7703/0.7703	74
TIME	0.9521/0.8872/0.9185	470

(*continued*)

Table 2. (*continued*)

Rule-Based Prediction Model		
STREET	0.9603/0.8430/0.8978	344
CITY	0.9694/0.9357/0.9523	373
ZIP	0.9915/0.9887/0.9901	353
DEPARTMENT	0.8690/0.8234/0.8456	419
STATE	0.9909/0.9849/0.9879	332
DURATION	0.7500/0.5000/0.6000	12
AGE	0.9167/0.8627/0.8889	51
LOCATION-OTHER	1.0000/0.5000/0.6667	6
SET	1.0000/0.6000/0.7500	5
PHONE	1.0000/1.0000/1.0000	1
Micro-avg. F	0.9166/0.9077/0.9121	13007
Macro-avg. F	0.9189/0.8406/0.8780	13007
Task 2		
Coding Type	*Precision/Recall/F1-score*	*Support*
DATE	0.8210/0.7552/0.7867	2459
TIME	0.7482/0.6638/0.7035	470
DURATION	1.0000/0.5000/0.6667	12
SET	1.0000/0.6000/0.7500	5
Micro-avg. F	0.8103/0.7393/0.7732	2946
Macro-avg. F	0.8923/0.6298/0.7384	2946

4 Discussion

Our findings elucidate the pivotal role of rule-based methods in augmenting clinical entity recognition, particularly in mitigating the challenges posed by data scarcity for specific entities. Despite the high precision demonstrated by rule-based models, opportunities for further refinement in recall metrics have been identified, underscoring the need for ongoing adjustments to enhance rule-based approaches for reliable performance across diverse datasets. A noteworthy challenge encountered, as illustrated in Fig. 2, was the model's inclination to generate abbreviated responses to detailed prompts. This limitation was attributed to its predefined word generation cap. In response, we recalibrated the model to accept a higher maximum word count, elevating it to 100 words. This strategy aims to substantially mitigate instances of truncated content. This adjustment is anticipated to markedly improve the thoroughness of the model's output, enhancing both the accuracy and reliability of the text generation process. This enhancement is particularly vital for complex tasks where the depth and completeness of the content are essential for accurate data interpretation and application.

Prevailing research substantiates the integration of rule-based methods as a means to enhance the de-identification process's efficiency, underscoring the significant potential these methods hold in addressing de-identification challenges [24–26]. Our research leverages these insights, highlighting the practical application of rule-based methods within an AI-driven framework to navigate the complex challenges of healthcare data privacy. Through the application of rule-based enhancements alongside the T5-Efficient-BASE-DL2 model, we have achieved significant improvements in de-identifying sensitive health information (SHI) and normalizing temporal information. In comparison with the OpenDeID pipeline [27], our study similarly emphasizes the critical importance of rule-based methods. However, our research extends the utility of these methods by specifically targeting efficiencies in identifying SHI and normalizing temporal data within a broader dataset. This demonstrates the adaptability and scalability of such hybrid approaches in handling diverse and complex datasets.

The remarkable F1 score achieved by the OpenDeID pipeline underscores the potential of transformer-based language models when integrated with rule-based systems [27]. Our findings complement this by illustrating that substantial improvements in de-identification performance are attainable even with different transformer models, suggesting the broad applicability and benefit of combining rule-based approaches with advanced machine-learning techniques across various models and datasets. Our future strategy involves a methodological refinement that incorporates complete texts as in-its, aiming to enhance the model's contextual understanding and, thus, its recognition accuracy. Efforts will also be directed towards addressing challenges with minority entity types, such as COUNTRY and LOCATION-OTHER, and the normalization of temporal data, like DURATION. By curating a dataset specifically tailored for these entity types and temporal information, we aim to advance our methodologies toward achieving heightened accuracy and efficiency in confidential healthcare data processing.

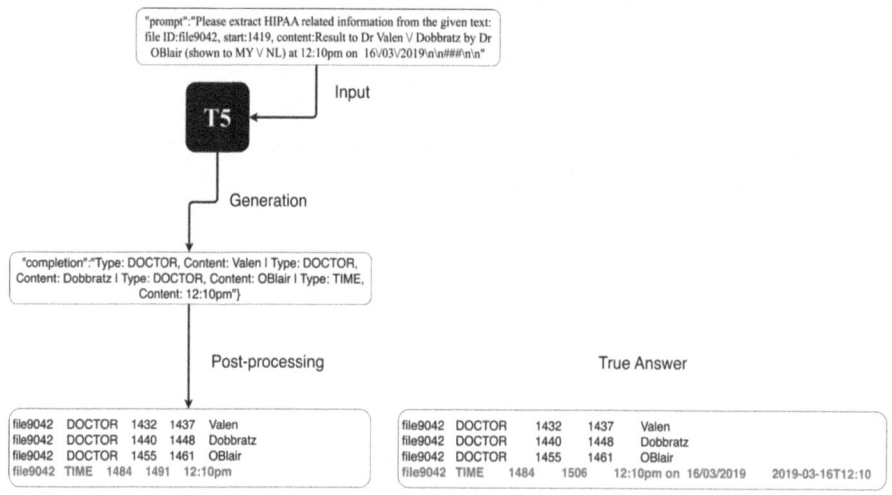

Fig. 2. Example of model-generated outcome with standard.

5   Conclusion

The findings of this research represent a significant advancement in the automated de-identification of patient data. By fine-tuning the T5-Efficient-BASE-DL2 model and implementing Exact Match metrics for evaluation, we have enhanced the model's ability to process sensitive information with high precision. In particular, the integration of extensive rule-based post-processing has markedly improved the recognition of nuanced entity types, such as phone numbers and geographic locations. Through the application of regular expressions and the Pycountry library [22], combined with a dedicated mapping system for temporal durations, we have successfully standardized the model's outputs in accordance with international norms. The results have shown notable improvements, with the macro-average F1-score for Task 1 increasing from 0.7787 to 0.9187 and for Task 2 from 0.4871 to 0.79, illustrating the significant enhancement in the model's capability to identify and normalize critical entities.

These advancements not only strengthen the security and efficiency of healthcare information management but also uphold individual privacy. It facilitates broader data access for research purposes without compromising the confidentiality of personal health information. Moving forward, we plan to further refine our rule-based processing techniques, augment our datasets with more diverse samples, and deepen our exploration of the interplay between machine learning algorithms and linguistic rule systems. Our ongoing work is set to drive progressive improvements in the field of data privacy and set new standards in the management of sensitive information.

Acknowledgments. This study did not receive any specific grant from funding agencies in the public, commercial, or not-for-profit sectors. We would like to acknowledge the Ministry of Education in Taiwan for sponsoring the Artificial Intelligence CUP 2023 - Privacy Protection and Medical Data Standardization Challenge. This initiative, part of the "Ministry of Education Artificial Intelligence Competition and Annotation Data Collection Project," provided a valuable context for this research. We are also grateful for the use of the OpenDeID corpus, which significantly facilitated our study we recognize the IW-DMRN workshop (https://www.sredhconsortium.org/sredh-competitions/sredhai-cup-2023/2024-iw-dmrn) and the SREDH Consortium (https://www.sredhconsortium.org/) for their valuable contributions to our research endeavors.

Disclosure of Interests.   The authors declare no competing interests relevant to the content of this article.

Data Availability.   In alignment with open science principles, the source code developed during this study is openly available on GitHub: https://github.com/nlptmu/ClinCaseCipher. This initiative supports Springer's advocacy for open science, encouraging the scientific community to reuse and expand upon our work.

Data Privacy Notice: The OpenDeID V2 corpus, the sole dataset used in this project, involves de-identified patient information to protect privacy, adhering to the HIPAA and other privacy standards. Due to the sensitive nature of this data and to maintain the highest level of confidentiality and data protection, the dataset itself is not publicly accessible. The code and computational notebooks we provide are designed for demonstration purposes, illustrating the approach and potential outcomes of our work. These materials are not directly executable without access to the OpenDeID v2 corpus, reflecting our commitment to patient privacy and data security.

References

1. Ge, Y., et al.: OpenAGI: when LLM meets domain experts. In: Oh, A., Neumann, T., Glober-son, A., Saenko, K., Hardt, M., Levine, S. (eds.) Advances in Neural Information Processing Systems, pp. 5539– 5568. Curran Associates, Inc. (2023)
2. Zharovskikh, A.: ChatGPT Large Language Model Explained. https://indat-alabs.com/blog/chatgpt-large-language-model. Accessed 4 Mar 2024
3. Mearian, L.: What are LLMs, and how are they used in generative AI? https://www.com-put erworld.com/article/3697649/what-are-large-language-models-and-how-are-they-used- in-generative-ai.html. Accessed 4 Feb 2024
4. Cascella, M., Montomoli, J., Bellini, V., Bignami, E.: Evaluating the feasibility of ChatGPT in healthcare: an analysis of multiple clinical and research scenarios. J. Med. Syst. **47**, 33 (2023). https://doi.org/10.1007/s10916-023-01925-4
5. Health, T.L.D.: Large language models: a new chapter in digital health. Lancet Digit. Health **6**, e1 (2024). https://doi.org/10.1016/S2589-7500(23)00254-6
6. Neel, S., Chang, P.: Privacy issues in large language models: a survey (2024). http://arxiv.org/abs/2312.06717
7. Fernández-Alemán, J.L., Señor, I.C., Lozoya, P.Á.O., Toval, A.: Security and privacy in electronic health records: a systematic literature review. J. Biomed. Inform. **46**, 541–562 (2013). https://doi.org/10.1016/j.jbi.2012.12.003
8. SREDH - SREDH/AI Cup 2023. https://www.sredhconsortium.org/sredh-competitions/sre dhai-cup-2023. Accessed 4 Mar 2024
9. SREDH - 2024 IW-DMRN. https://www.sredhconsortium.org/sredh-competitions/sredhai-cup-2023/2024-iw-dmrn. Accessed 4 Mar 2024
10. AI CUP 2023|AI CUP. https://www.aicup.tw/copy-of-ai-cup-2023-1 Accessed 4 Mar 2024
11. ISLAB: CodaLab – Competition. https://codalab.lisn.upsaclay.fr/competitions/15425. Accessed 4 Mar 2024
12. Chen, A., Jonnagaddala, J., Nekkantti, C., Liaw, S.-T.: Generation of surrogates for de-identification of electronic health records. In: MEDINFO 2019: Health and Wellbeing e-Networks for All, pp. 70–73. IOS Press (2019)
13. Alla, N.L.V., Chen, A., Batongbacal, S., Nekkantti, C., Dai, H.-J., Jonnagaddala, J.: Cohort selection for construction of a clinical natural language processing corpus. Comput. Methods Prog. Biomed. Update **1**, 100024 (2021). https://doi.org/10.1016/j.cmpbup.2021.100024
14. Jonnagaddala, J., Chen, A., Batongbacal, S., Nekkantti, C.: The OpenDeID corpus for patient de-identification. Sci. Rep. **11**, 19973 (2021). https://doi.org/10.1038/s41598-021-99554-9
15. Herold, R., Beaver, K.: The Practical Guide to HIPAA Privacy and Security Compliance. CRC Press (2003)
16. Lhoest, Q., et al.: Datasets: a community library for natural language processing. arXiv preprint arXiv:2109.02846 (2021)
17. Raffel, C., et al.: Exploring the limits of transfer learning with a unified text-to-text transformer (2023). http://arxiv.org/abs/1910.10683. https://doi.org/10.48550/arXiv.1910.10683
18. Tay, Y., et al.: Scale efficiently: insights from pre-training and fine-tuning transformers. arXiv preprint arXiv:2109.10686 (2021)
19. Lin, C.-Y.: Rouge: a package for automatic evaluation of summaries. Presented at the Text summarization branches out (2004)
20. Pedregosa, F., et al.: Scikit-learn: machine learning in Python. J. Mach. Learn. Res. **12**, 2825–2830 (2011)
21. Wolf, T., et al.: Transformers: state-of-the-art natural language processing. In: Proceedings of the 2020 Conference on Empirical Methods in Natural Language Processing: System Demon-strations, pp. 38–45. Association for Computational Linguistics, Online (2020). https://doi.org/10.18653/v1/2020.emnlp-demos.6

22. Theune, C.: pycountry (2022). https://github.com/pycountry/pycountry
23. Wolf, M., Wicksteed, C.: Date and time formats. W3C NOTE-datetime-19980827, August 1998
24. Zhao, Z., Yang, M., Tang, B., Zhao, T.: Re-examination of rule-based methods in deidentification of electronic health records: algorithm development and validation. JMIR Med. Inform. **8**, e17622 (2020). https://doi.org/10.2196/17622
25. Kim, Y., Heider, P., Meystre, S.: Ensemble-based methods to improve de-identification of electronic health record narratives. AMIA Annu. Symp. Proc. **2018**, 663–672 (2018)
26. Trienes, J., Trieschnigg, D., Seifert, C., Hiemstra, D.: Comparing rule-based, feature- based and deep neural methods for de-identification of Dutch medical records (2020). http://arxiv.org/abs/2001.05714
27. Liu, J., et al.: OpenDeID pipeline for unstructured electronic health record text notes based on rules and transformers: deidentification algorithm development and validation study. J. Med. Internet Res. **25**, e48145 (2023). https://doi.org/10.2196/48145
28. Mir, T.H., et al.: Deidentification and temporal normalization of the electronic health record notes using large language models: the SREDH/AI-Cup 2023 deidentification competition. In: 2024 International Workshop on Deidentification of Electronic Medical Record Notes. Kaohsiung, Taiwan. Springer, Cham (2024)

A Comparative Study of GPT3.5 Fine Tuning and Rule-Based Approaches for De-identification and Normalization of Sensitive Health Information in Electronic Medical Record Notes

Zi-Rui Zhao[1,5] ⓘ, Po-Chen Chou[2,5] ⓘ, Tatheer Hussain Mir[1,2,5] ⓘ,
and Hong-Jie Dai[1,3,4,5(✉)] ⓘ

[1] Intelligent System Laboratory, Department of Electrical Engineering, College of Electrical Engineering and Computer Science, National Kaohsiung University of Science and Technology, No. 415, Jiangong Road, Sanmin District, Kaohsiung City 807618, Taiwan
{c110154237,hjdai}@nkust.edu.tw
[2] Department of Electrical Engineering, College of Electrical Engineering and Computer Science, National Kaohsiung University of Science and Technology, No. 415, Jiangong Road, Sanmin District, Kaohsiung City 807618, Taiwan
c110154203@nkust.edu.tw
[3] National Institute of Cancer Research, National Health Research Institutes, Tainan, Taiwan
[4] Center for Big Data Research, Kaohsiung Medical University, No. 100, Shihcyuan 1st Road, Sanmin District, Kaohsiung City 807, Taiwan
[5] School of Post-Baccalaureate Medicine, College of Medicine, Kaohsiung Medical University, Shihcyuan 1st Road, Sanmin District, Kaohsiung City 807, Taiwan

Abstract. Electronic Medical Records (EMR) systems enhance medical care efficiency through the consolidation of data analysis, bolstering patient safety, and diminishing storage expenses. Nevertheless, EMR text notes have the potential to reveal sensitive personal information, necessitating the implementation of robust security measures to protect sensitive information. Furthermore, the representation of temporal data in EMR text notes differs among institutions, which has a direct effect on the precision and dependability of time data analysis. In light of this, the AI CUP 2023-Privacy Protection and Standardization of Electronic Medical Record competition released a dataset annotated with sensitive health information (SHI) and normalized temporal information. We participated in the challenge and conducted a comparative study to assess the efficacy of two approaches, the fine-tuning of a large language model based on the chat generative pre-trained transformer (ChatGPT) and the rule-based approach, in enhancing the privacy of clinical texts. For the ChatGPT-based approach, we fine-tuned the gpt-3.5-turbo-1106 model to efficiently identify labels. In contrast, in the rule-based approach, we compiled several rules along with dictionaries collected from the internet and released datasets for recognizing and normalizing SHIs. The ChatGPT-based approach achieved macro-F-scores of 0.752 and 0.799 for SHI recognition and temporal information normalization, while the rule-based approach highlighted 0.866 and 0.869 for the two respective subtasks, exhibiting lower latency and power consumption. Both strategies exhibited their respective benefits in tackling privacy concerns. The implementations of both rule-based and ChatGPT-based methods can be accessed through the following GitHub repositories: https://github.com/zhao-rui-NB/worker_intelligence_ner (rule-based method) and https://github.com/Chou-po-chen/openai_ner (ChatGPT-based method).

Keywords: Natural language processing · de-identification · sensitive health information · temporal information normalization · ChatGPT · rule-based approach

1 Introduction

The Health Insurance Portability and Accountability Act (HIPAA) in 1996 [1] led to the adoption of electronic health records (EHRs) systems by medical establishments worldwide. EHRs have become essential for clinical data analysis and have facilitated the digitization of patient records into electronic medical records (EMRs). As the digitization of healthcare information advances, the protection of patient privacy becomes increasingly crucial. HIPAA and other regulations mandate healthcare institutions to ensure the privacy and security of patient information. To effectively use EMR text notes for secondary data analysis, patient privacy data needs to be removed, but the challenge lies in adapting to EMR text notes in different languages and cultural backgrounds. Temporal description in EMR text notes is another issue, making it challenging to analyze and compare time information.

To properly utilize EMR text notes, the 2024 International workshop on de-identification of EMR notes [2, 3] organized the AI CUP 2023 Privacy Protection and Standardization of Electronic Medical Record (PS-EMR) [4, 5] competition which released a de-identification and temporal normalization corpus aiming to address the challenges of handling patient privacy data in EMR text notes, safeguarding patient privacy rights, and enhancing the security of healthcare information systems. In response to the PS-EMR competition, we earnestly dedicated our endeavors to confront and engage with the challenges by using two approaches: the large language model (LLM) fine-tuning method based on the chat generative pre-trained transformer (ChatGPT) [6, 7] and the rule-based approach. The selection of the two methodologies was based on their complementary strengths. LLM fine-tuning is the recent advancement of natural language processing (NLP), which leverages the power of pre-trained language models, specifically tailored for downstream tasks. By fine-tuning ChatGPT, we could establish a strong baseline which could utilize LLMs' ability to understand and generate desired output text. This approach is particularly effective in scenarios where the data is diverse and nuanced, enabling the system to learn and adapt to various patterns and contexts. On the other hand, the rule-based approach offers a structured and deterministic way of handling information [8]. By defining explicit rules and patterns, we can ensure precise control over how data is processed and interpreted. This method is advantageous when dealing with texts characterized with consistent patterns in their descriptions or tasks with specific guidelines or regulations that need to be strictly adhered to. Additionally, rule-based systems are often more interpretable and transparent, which is crucial in medical applications where the decisions made by the system need to be justified and understood by healthcare professionals.

We aimed to harness the strengths of each method to provide suggestions for developing robust de-identification systems capable of effectively addressing the challenges of SHI recognition and temporal information normalization. To do this, we studied the performance of these two approaches on the corpus released by the PS-EMR challenge. Our goal was to capitalize on the strengths of each method.

2 Methodology

2.1 Dataset

The biobank of the Health Science Alliance (HSA), which is located at the Lowy Cancer Research Center of the University of New South Wales (UNSW) in Australia, serves as the competition's primary source of data [9, 10]. The OpenDeID v2 corpus is comprised of 3,244 pathology reports, which are divided into 1,734 training sets, 560 validation sets, and 950 testing sets. As a result of its characteristics, the OpenDeID corpus is a very useful resource for building and enhancing automated patient de-identification systems. Additionally, the corpus is the first gold-standard corpus that was generated specifically for this purpose and is headquartered anywhere in Australia. A further point to consider is that the corpus has been annotated with 38,414 Sensitive Health Information (SHI) items that have good accuracy and deviation ratings in terms of inter-annotator agreement. One of the distinctive features is that it has been carefully annotated with surrogate information, which helps to ensure that patient information remains anonymous. Because of its relevance and variety, this dataset is a very significant resource for research activities that are being conducted in the healthcare industry. Table 1 summarizes the SHI categories defined in the guideline.

Table 1. HSA Study SHI Corpus - Annotation guidelines 5.0

SHI category	Sub-category
NAME	PATIENT, DOCTOR, USERNAME
PROFESSION	(none)
LOCATION	ROOM, DEPARTMENT, HOSPITAL, ORGANIZATION, STREET, CITY, STATE, COUNTRY, ZIP, OTHER
AGE	(none)
DATE	DATE, TIME, DURATION, SET
CONTACT	PHONE, FAX, EMAIL, URL, IPADDRESS
IDs	SOCIAL SECURITY NUMBER, MEDICAL RECORD NUMBER, HEALTH PLAN NUMBER, ACCOUNT NUMBER, LICENSE NUMBER, VEHICLE ID, DEVICE ID, BIOMETRIC ID, ID NUMBER
OTHER	None

2.2 ChatGPT-Based Method

The utilization of a ChatGPT-based approach involves employing OpenAI's ChatGPT model to perform natural language processing tasks. This model is trained using a wide range of text to produce responses that resemble humans, participating in conversations, and completing tasks related to generating text. It is utilized for tasks such as text summarization, question answering, translation, and dialogue systems. ChatGPT possesses remarkable flexibility and adaptability, rendering it a potent instrument for natural language processing and artificial intelligence applications. Researchers and developers utilize it for constructing conversational agents, chatbots, recommendation systems, and content-generation tools.

For the ChatGPT-based method, the study used the gpt-3.5-turbo-1106 model released by OpenAI [11]. The implementation's user-friendliness and impressive language processing capabilities were the deciding factors. The technique comprised three essential phases, as described in the following subsections.

Training Dataset Construction. For each pathology report and the corresponding annotation file, we applied sentence segmentation to segment sentences [12] and associated each report with the corresponding annotations sorted according to their start indexes to generate the pre-processed dataset format shown in Table 2.

Table 2. Dataset preprocessed for the ChatGPT model. The content shown within the last column is optional.

Pathology Report	Text in the referenced report	Annotation Type	Temporal Annotation
Tokenized and truncated report1	SHI annotated text1	SHI type1	=> Normalization value1
	SHI annotated text2	SHI type2	=> Normalization value1
	...		
Tokenized and truncated report2	SHI annotated text1	SHI type1	=> Normalization value1
	SHI annotated text2	SHI type2	=> Normalization value1
	...		
...	...	...	...

Afterward, the sentences whose start index exceeded the context length limitation (4,096 tokens) of the ChatGPT API were removed, resulting in a truncated report as shown in Table. The outcomes are stored in the respective files. As depicted in Table 2, the generated data includes annotated content, the SHIs, and the normalized values (for DATE, TIME, DURATION, and SET labels).

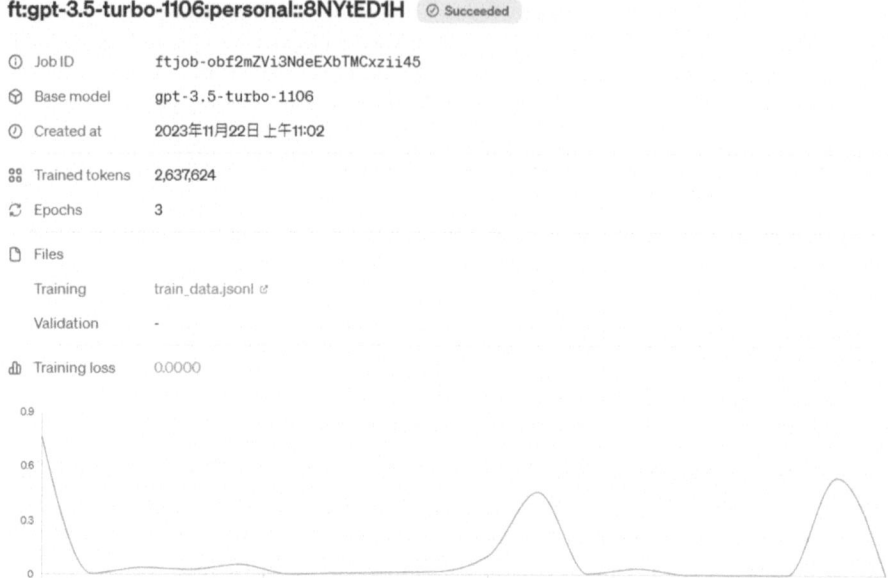

Fig. 1. The initial trained model entitled "ft:gpt-3.5-turbo-1106:personal::8NYtED1H".

Instruction-Based Fine-Tuning. We utilized the closed-source pre-trained large language model (LLM) model, gpt-3.5-turbo-1106, and applied instruction fine-tuning [12] to develop the system for the AICUP PS-EMR. The focus was on improving the model's capability for recognizing SHIs defined in Table 1, especially in situations where there are limited labeled examples. For the ChatGPT development environment, we used the OpenAI python library (version 1.3.3) developed by Open AI, and the tiktoken module (version 0.5.1), which provides the fast Byte-Pair Encoding (BPE) tokenizer for use with OpenAI's models. The gpt-3.5-turbo-1106 model was used as the pre-training model, which is a paid model with token-based pricing. The pricing information consists of $0.0080 per 1000 tokens for training, $0.0030 per 1000 tokens for input, and $0.0060 per 1000 tokens for output.

Following the specification of ChatGPT API, the message of a prompt contains three roles: system, user, and assistance. Every role is characterized by a name (role) and its corresponding material (content). The "system" role is denoted by the system command as follows.

Retrieve SHI (DOCTOR, DATE, IDNUM, MEDICALRECORD, PATIENT, HOSPITAL, TIME, DEPARTMENT, CITY, ZIP, STREET, STATE, AGE, ORGANIZATION, DURATION, PHONE, URL, LOCATION-OTHER, SET, COUNTRY, ROOM) and standardize time data.

The truncated report that needs to be de-identified is associated with the "user" role, which is the first column of Table 1. The "assistant" role is responsible for generating the desired output, which is the second, third, and fourth columns of Table 1. To begin, we train the original GPT-3.5 Turbo 1106 model by selecting 600 sentences at random from the initial compiled training dataset. This is illustrated in Fig. 1. Following the production of the model, the identifier (8NYtED1H) is assigned to it, and the total number of training tokens is 2,637,624.

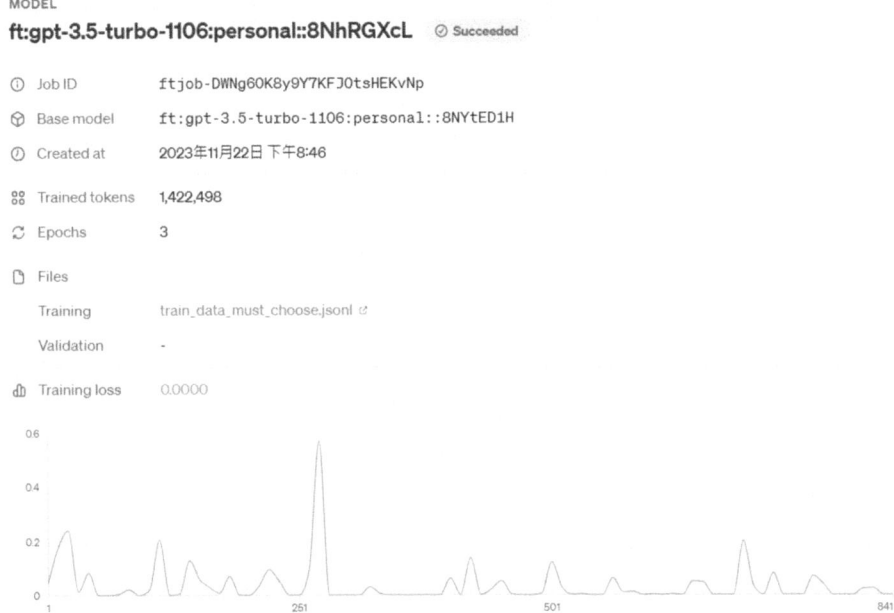

Fig. 2. The last trained model entitled "ft:gpt-3.5-turbo-1106:personal::8NhRGXcL"

After the initial training, the model is provided with the remaining training set. To maintain the training costs effectively, we adopted a sampling approach from the entire training set. Specifically, we selected 281 reports that contain SHI types such as age, organization, duration, phone, URL, location-other, set, and country, all of which occurred fewer than 150 times. The initial fine-tuned model then undergoes a second training process on the additional sampled training data. Since the competition score is calculated by taking the average of individual scores, the purpose of this second training is to improve the categories that have fewer occurrences, ultimately leading to an overall improvement in the total F-score. As shown in Fig. 2, the new model obtained the identifier 8NhRGXcL following the completion of the second training phase, during which it processed a total of 1,422,498 training tokens. As it is proprietary, the particular parameter setup is not revealed. The trained model is maintained on the OpenAI server, which can be accessed by using its ID with the official Open API commands.

In addition to GPT3.5-Turbo, we also consider training a model based on the smaller model, babbage-002, which has a lower cost of training. Similar to the above procedure, we utilized all the training sets to train the babbage-002 model, as shown in Fig. 3. Nevertheless, the trained model generates numerous duplicate outputs. Consequently, we opted to stick with gpt-3.5-turbo-1106 as the final model during the competition.

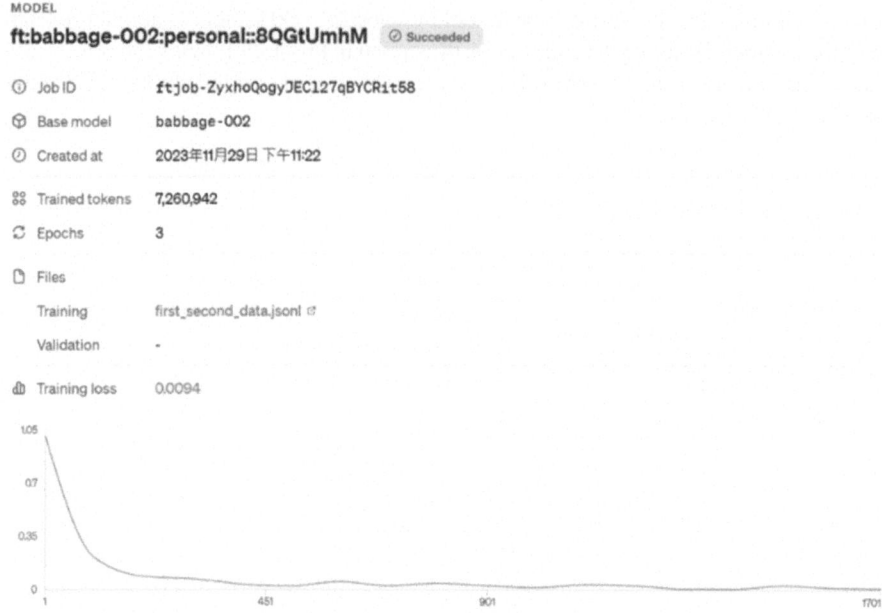

Fig. 3. The trained babbage-002 model entitled "ft:babbage-002-personal:BQGtmhM"

Post-processing. The procedure entails utilizing the string-matching method to determine the start and end indexes of the recognized SHIs from the given pathology report based on the generated output of the fine-tuned model. In some cases, the ChatGPT model might exhibit hallucinations [16]; for example, there are SHI types generating by the fine-tuned LLM which are not defined in Table 1. To address this, we compiled a lexicon that encompasses all of the SHI-type names, which is utilized to validate whether the ChatGPT-generated SHI types belong to the desired categories defined in Table 1. This post-processing step can eliminate unexpected SHI types, such as "DECADE" shown in Table 3, which may be generated by the fine-tuned model.

Table 3. Examples of hallucination produced by the fine-tuned ChatGPT model. The unmatched SHIs are highlighted in bold.

Input Text	Generated Output
Additional history from Dr BAILEY Bacayo, testicular tumour forty years ago. Recent biopsy at NARACOORTE–sarcoma	**DECADE:** forty years => P40Y

2.3 Rule Based Method

Rule-based machine learning methods use predefined rules or logical conditions derived from domain knowledge or expert insights to classify or predict outcomes for new data instances. These methods do not learn relationships between input features and output labels through optimization algorithms like gradient descent or backpropagation. They offer advantages like transparency, interpretability, and ease of understanding but may struggle to capture complex relationships or handle noise and variability effectively in high-dimensional or unstructured data. Examples include decision trees, expert systems, and production rule systems.

In this approach, we developed rules represented by regular expressions, providing an alternative to machine learning-based methods. We developed an analysis tool to provide a comparative function to facilitate the assessment of disparities between the gold training set and the predicted results based on the compiled rules. The tool classifies the disparities as false positive and false negative instances for further manually examining, establishing, or altering our patterns until the desired score is attained. We dedicated one month to meticulous manual examination of the dataset. Our comprehensive analysis led us to the conclude that capturing SHIs mentioned in the pathology reports and normalizing temporal information could be effectively achieved by employing three distinct types of rules, summarized in Table 4.

Table 4. The rule categories developed for the AICUP-PS-MER.

Rule Category	SHI Category
Pattern	ZIP, URL, TIME, STREET, PHONE, STATE, SET, PATIENT, AGE, MEDICALRECORD, LOCATION_OTHER, IDNUM, DOCTOR, DURATION, DATE, COUNTRY
Dictionary	ORGANIZATION, CITY
Hybrid	HOSPITAL DEPARTMENT

Pattern-Matching. We observed that certain SHIs mentioned in the corpus have strong regular patterns. Therefore, we used Python "re" module to compile regular expressions for pattern matching. We developed the LABEL_Finder module, which contains the Finder class and subclasses for pattern-based methods such as AGE_Finder. The classes defined in the module apply precise patterns to effectively detect and extract the desired SHIs. A few examples of the developed patterns for each SHI category are depicted in Table 5.

Dictionary-Based Matching. Dictionary-based matching is a natural language processing technique used for identifying matches in text using a predefined dictionary. It is commonly used in tasks like named entity recognition, keyword extraction, sentiment analysis, spell checking, and information retrieval. However, its effectiveness may be limited by the dictionary's coverage and ability to handle word variations.

Table 5. The example patterns defined for the SHI types belong to the pattern rule category. SHIs marked with an asterisk at end indicate that they are recognized using the hybrid approach.

SHI Category	SHI Category
AGE	(?i)(\d{2,3}) *years *old (?i)(\d{2,3}) *yrs *old (?i)\n(\d{2,3}) ?F\b'. ##(?!\.) (?i)\n(\d{2,3}) ?M\b'. ##(?!\.) (?i)\bage *(\d{2,3})\b' (?i)(\d{2,3}) *female' cancer at (\d{2,3})s?\b
DATE	(\d{1,2}[/.]\d{1,2}[/.]\d{2,5}) DateOfBirthn(\d{8})\n ((?:\d{1,2})\s*)?(?:January|February|March|April|May|June|July|August|September|October|November|December) \d{2,5})
DEPARTMENT*	Location: *([A-Za-z0-9'] + (?: + [A-Za-z0-9'] +)*) *-
DOCTOR	(?i)[\n]DR + ([A-Za-z\-]* ?[A-Za-z\-]* ?[A-Za-z\-]*)[\n] (?i:Reported ?by ?Dr *)([A-Z][A-Za-z\-]*(?: [A-Z][A-Za-z\-]*)?(?: [A-Z][A-Za-z\-]*)?) \(([A-Z][A-Z])/ ?ec [Dd]\[Rr]\.? ([A-Z][A-Za-z\-]*(?: [A-Z][A-Za-z\-]*)?(?: [A-Z][A-Za-z\-]*)?) Prof\.? ([A-Z][A-Za-z\-]*(?: [A-Z]['A-Za-z\-]*)?(?: [A-Z]['A-Za-z\-]*)?) (?i:Reported ?by)([A-Z][A-Za-z\-]*(?: [A-Z][A-Za-z\-]*)?(?: [A-Z][A-Za-z\-]*)?) (?i)\n ?MICROSCOPIC *:? *(Reported by Dr ([A-Z][A-Za-z\-]*(?: [A-Z][A-Za-z\-]*)?(?: [A-Z][A-Za-z\-]*)?) *\) (?i)\n ?MICROSCOPIC *:? *(Reported by ([A-Z][A-Za-z\-]*(?: [A-Z][A-Za-z\-]*)?(?: [A-Z][A-Za-z\-]*)?) *\) (?i)\n ?MICROSCOPIC *:? *(Reported ([A-Z][A-Za-z\-]*(?: [A-Z][A-Za-z\-]*)?(?: [A-Z][A-Za-z\-]*)?) *\) (?i)\n ?MICROSCOPIC *:? *((([A-Z][A-Za-z\-]*(?: [A-Z][A-Za-z\-]*)?(?: [A-Z][A-Za-z\-]*)?) *\)

(continued)

Table 5. (*continued*)

SHI Category	SHI Category
DURATION	(?i)(\d{1,3}(?:-\d{1,3})? *years)\b(?! old)
	(?i)(\d{1,3}(?:-\d{1,3})? *year)\b(?! old)
	(?i)(\d{1,3}(?:-\d{1,3})? *yrs)\b(?! old)
	(?i)(\d{1,3}(?:-\d{1,3})? *yr\b)\b(?! old)
	(?i)(\d{1,3}(?:-\d{1,3})? *weeks)
	(?i)(\d{1,3}(?:-\d{1,3})? *week)
	(?i)(\d{1,3}(?:-\d{1,3})? *wks)
	(?i)(\d{1,3}(?:-\d{1,3})? *months)
	(?i)(\d{1,3}(?:-\d{1,3})? *month)
	(?i)(\d{1,3}(?:-\d{1,3})? *day)
	(?i)(\d{1,3}(?:-\d{1,3})? *days)
HOSPITAL*	Site_name: *(.{6,})\n
	Site_name: *(.{6,})\n
	Site: *([A-Z] + HOSPITAL(?:[A-Z])*)\n
	\nSiteName\n(. +)\n
	Location:.*?- *(?:OPERATIVE UNIT|OPERATIVEUNIT)? + [\-]*([' &A-Z\-\(\)] +) *\n

(*continued*)

Table 5. (*continued*)

SHI Category	SHI Category
IDNUM	(?i)lab *no:? *(\w +),?,?(\w +)?
	(?i)episode *no:? *(\w +),?,?(\w +)?
	(?i)SPR *no:? *(\w +),?,?(\w +)?
	\A(?:.* + \n){1,9}?(\w{7,8}),?(\w{7,8})?\n
	\nSPRIDn(\w{7,11})\n
	Pathology Report(\w{7,11})\b (?!No:)
	Pathology Report(\w{7,11})\b,(\w{7,11})
	Laboratory Number: (\w{7,11})\b,(\w{7,11}) report (\w{7,11})\b
	[\n](?P < id > \w{8,11}), ?(?P = id)
	SPRTextn(\w{7,11})\b
	SPRTextn\w{7,11}\((\w{7,11})\)
	Pathology number ([A-Z\d-]{7,12})\b \(([A-Z\d]{7,11})\)
LOCATION_OTHER	(?i)(PL.? ?OL.? *BOX *\d +)
MEDICALRECORD	(?i)MRN no: *(\w +)\n?
	(\d{6,8}\.[A-Z][A-Z][A-Z])
	\nMRN\n(\d +)\n
	\nMRN: ?(\d + (?:\.[A-Z][A-Z][A-Z])?)
PATIENT	(? < !\[Dd][Rr])(\b[A-Z][A-Za-z]*(?: ?, [A-Z][A-Za-z\-']*,?)(?: ?: [A-Z][A-Za-z\-']*)?)?(?: [A-Z][A-Za-z\-']*)? *\n
	\nFirstName\n([A-Z][A-Za-z]*)\n
	\nMiddleName\n([A-Z][A-Za-z\-]*)\n
	\nLastName\n([A-Z][A-Za-z\-]*)\n

(*continued*)

Table 5. (*continued*)

SHI Category	SHI Category
PHONE	\(\(\d\d\d\d \d\d\d\d\)\) (?i)Phone: *(\d\d\d\d \d\d\d\d) (?i)Tel: *(\d\d\d\d \d\d\d\d) \(\(\d\d\d\d\d\d\d\d\)\)
SET	\b(once)\b \b(twice)\b
STREET	Lab ?No:.*\n([A-Z][A-Za-z\.]*(?: + [A-Zo][a-z\.]*)*)\b
TIME	(\d\d\d\d-\d\d-\d\d\d\d:\d\d:\d\d) (?i)(\d{1,2}[\A.]\d{1,2})[\A.]\d{2,4} *(?:at|and|on|@)? *\d{1,2}[\A.]\d{1,2}(?: ?(?:am|pm|hrs|hr))?)? (?i)(\d{1,2}[\A.]\d{1,2})[\A.]\d{2,4} *(?:on the|on|at|and|,|@)? *\d[1](?: ?(?:am|pm|hrs|hr))?)? (?i)(\d{1,2}[\A.]\d{1,2})[\A.]\d{2,4} + (?:on the|on|at|and|,|@)? *\d{1,2}(?: ?(?:am|pm|hrs|hr))?)? (\d{1,2}[:\.]\d{1,2} ?(?:am|pm|hr|hrs)? *(?:on the|on|at|and|,|@)? *\d{1,2}[\A.]\d{1,2}[\A.]\d{2,4})
URL	(http[s]?:/[^\n^"^\(^\)] +)
ZIP	Lab ?No:(?:.*\n){1,3}?. + [A-Z](\d\d\d\d)\n

We observed that the SHIs belonging to ORGANIZATION cannot produce any regular expressions. Therefore, we relied on the dictionary look-up method using two pre-made lexicons:

- Keyword filtering matching (ORGANIZATION_dict_V3.txt): This lexicon is relatively strict in its matching criteria. If the length of a dictionary entry is less than 9 characters, the result string must contain predefined *ORG_KEYWORD* (such as Services, Service, USA, Co, Ltd) to be considered a valid match. The lexicon file contains the names of the 500 most prominent companies in the world, sourced from internet searches [17]. We processed the ORGANIZATION-related terms collected from the internet by removing text following commas, and spaces at the beginning and end, eliminating periods at the end of each line, and then splitting the words. In addition, we expanded the lexicon by eliminating designated terms such as Co, Ltd, Inc, Corp, Corporation, Company, and Group, and substituting ampersands with the word "and" to compile the final lexicon file.
- Basic matching (ORGANIZATION_force_dict.txt): In this case, any match with entries in the dictionary is directly considered as an ORGANIZATION SHI. The lexicon file *ORGANIZATION_force_dict.txt* contain the terms extracted from the training set annotated as ORGANIZATION-related SHIs. We also follow the same procedure to compile the lexicon file, *DEPARTMENT_force_dict.txt*, for the DEPARTMENT-related SHIs.

Hybrid Matching. Hybrid matching is an approach that integrates various matching techniques to improve the precision and efficiency of information retrieval and recommendation systems. The system utilizes both conventional keyword-based matching and sophisticated methods such as semantic matching and machine learning-based models. Recommendation systems employ a combination of collaborative filtering, content-based filtering, and knowledge-based methods. The objective is to capitalize on the advantages of various techniques while minimizing their limitations, leading to enhanced performance in these tasks.

For HOSPITAL and DEPARTMENT-related SHI types, we applied hybrid matching approaches. Take HOSPITAL as an example. In addition to using regular expression patterns shown in Table 3, we conducted specific searches for strings that are all in uppercase in brackets and contain one of the following keywords: SERVICE, HEALTH, HOSPITAL, CENTRE, and CAMPUS, which will most likely refer to HOSPITAL-related SHIs.

For the recognition of DEPARTMENT-related SHIs, our approach involves a combination of techniques. We first utilize the patterns outlined in Table 4 as well as the dictionary (DEPARTMENT_force_dict.txt) for a maximum matching search. Longer strings are given priority in this matching process. On the other hand, for the recognition of CITY-related SHIs, our strategy begins with pattern-based identification. We employ the *STATE_Finder* to initially locate the sentence where STATE-related SHIs are mentioned. Our analysis suggests that in this particular context, there is a high probability of finding the city name appearing before the STATE. Thus, we then extract all capitalized characters preceding the recognized STATE-related SHIs as a CITY SHI. In cases where no sentence contains STATE-related SHIs, the *ZIP_Finder* will be used instead of

STATE_Finde r and the above process is repeated in an attempt to identify CITY-related SHIs.

After all of the developed pattern-based modules have identified candidate SHIs, a post-processing module is employed to remove duplicate and overlapping candidates. Overlapping situations mostly occur for COUNTRY-related SHIs. Our analysis suggests that sometimes the correct HOSPITAL names also contain names of COUNTRY, so if there is an overlap between HOSPITAL-related and COUNTRY-related SHIs, the COUNTRY-related SHIs are deleted first.

Rule-Based Normalization Method. Rule-based normalization is a method in natural language processing that applies predetermined rules or patterns, such as case normalization, tokenization, and stemming, to standardize data. While these methods are easily understood and can be tailored to specific needs, they may not adequately address all possible variations or exceptional situations.

For the temporal normalization task, the target SHI types include DATE, TIME, DURATION, and SET. We specifically developed normalization rules for each date SHI pattern defined in Table 3. For example, one of the patterns used in the recognition of DATE-related SHIs is:

$$(\backslash d\{1,2\}[/\backslash.]\backslash d\{1,2\}[/\backslash.]\backslash d\{2,5\})$$

We developed patterns to group the strings to be extracted: When a match is found, we extract the text in the corresponding group and transform it into the standard ISO8601 format $(f'\{y\} - \{m : 02d\} - \{d : 02d\}')$ as defined in Table 6 to produce a normalized result.

Table 6. ISO8601 format output

SHI Category	format
DATE	f'{y}-{m:02d}-{d:02d}' f'{y}-{m:02d}'
TIME	f'{date}T{time}' f'{y}-{m:02d}-{d:02d}T{hh:02d}:{mm:02d}'
DURATION	f'P{num}Y' f'P{num}W' f'P{num}M' f'P{num}D'
SET	R1 R2

3   Results

During the final test period, we submitted two runs that were based on the methods that were originally proposed. The findings are presented in Table 7 and Fig. 4, respectively. The rule-based system demonstrates a superior F-score of 0.866, in contrast to the GPT-3.5 model, which achieved an F-score of 0.752 in subtask 1 (SHI recognition). This demonstrates that the rule-based system performs exceptionally well when it comes to identifying certain health indicators (SHIs). When it comes to subtask 2 (temporal information normalization), the rule-based system achieves an F-score of 0.869, while the GPT-3.5 turbo achieves a slightly better performance with an F-score of 0.799. https://github.com/zhao-rui-NB/worker_intelligence_ner is the location where you can gain access to the codes that were developed by our team, TEAM_3970. Particularly noteworthy is the fact that our rule-based approach achieved the highest possible ranking in both subtask 1 and subtask 2, which ultimately led to our overall first place ranking in the competition.

Table 7. The official precision (P), recall (R) and F-scores on the test set.

	GPT3.5 TURBO			Rule-based		
	P	R	F	P	R	F
Subtask 1(SHI recognition)	0.771	0.733	0.752	0.900	0.835	0.866
Subtask 2 (Temporal information normalization)	0.927	0.702	0.799	0.932	0.813	0.869

Based on the findings presented in Fig. 4, it becomes evident that the ChatGPT-based method significantly underperformed in comparison to the proposed rule-based method, particularly in recognizing PHONE and SET SHIs. This discrepancy might be attributed to the limited training instances available for both categories. Across all SHI types, except CITY, the rule-based method outperformed the ChatGPT-based method. This outcome suggests that the current implementation of compiled rules effectively captures the diverse writing styles present in the presented corpus.

4   Discussion

While rule-based methods achieve higher overall macro PRF scores, it's noteworthy that certain SHI types exhibit lower recall rates compared to GPT-based models. Specifically, rule-based methods achieved improved recall rates in types such as MEDI-CALRECORD, HOSPITAL, TIME, SET, LOCATION-OTHER and PHONE, alongside superior precision rates across nearly all SHI types. This advantage in precision stems from the method's reliance on predefined patterns observed frequently within the training set. However, this poses a problem as not all SHI Categories can rely on manually crafted perfect rules. Some SHI types, such as DOCTOR, PATIENT, and ORGANIZA-TION have significant variations or require understanding based on the context of the surrounding text. ChatGPT-based models achieved better recall for those types since they can learn and understand meanings and relationships among words in different contexts.

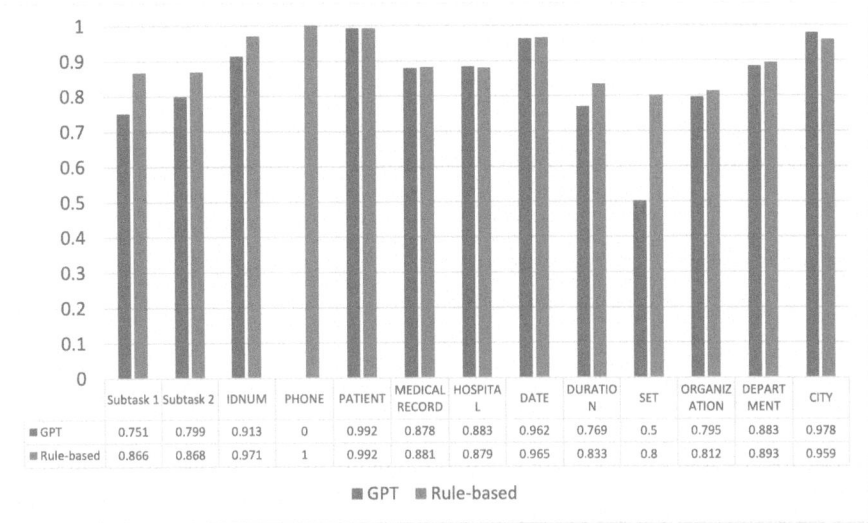

	Subtask 1	Subtask 2	IDNUM	PHONE	PATIENT	MEDICAL RECORD	HOSPITAL	DATE	DURATION	SET	ORGANIZATION	DEPARTMENT	CITY
GPT	0.751	0.799	0.913	0	0.992	0.878	0.883	0.962	0.769	0.5	0.795	0.883	0.978
Rule-based	0.866	0.868	0.971	1	0.992	0.881	0.879	0.965	0.833	0.8	0.812	0.893	0.959

■ GPT   ■ Rule-based

Fig. 4. Comparison between ChatGPT and rule-based methods.

Further comparative analysis was performed to assess the merits and drawbacks of both methods. The evaluation encompassed metrics including averaged macro-F-score, processing speed, and resource requirements. Detailed comparisons are given in Table 8. In terms of averaged performance, it is evident that the proposed handcrafted rule-based method surpasses ChatGPT for the task presented. ChatGPT relies on a complex neural network architecture to generate responses. The high computation demands of deep complex neural networks are reflected in the processing speed; the rule-based system, thanks to its straightforward pattern-matching methodology, exhibit significantly higher computational efficiency and faster processing speed compared to ChatGPT. Resource requirements reveal that ChatGPT, with its GPT-3.5 Turbo model, incurs higher costs and demands computational resources, whereas the rule-based approach is inherently less resource-intensive. However, the rule-based approach hinges on the manual examination of the dataset to develop patterns, revealing the trade-offs between data-driven and knowledge-intensive approaches. Overall, considering the training/prediction cost longer prediction time, and the high computational and energy demands associated with running LLMs like ChatGPT, the rule-based approach remains an appealing choice for the presented tasks consisting of SHI recognition and normalization.

Our study underscores the importance of considering not only performance metrics but also practical considerations such as computational efficiency and resource requirements when selecting an approach. While ChatGPT demonstrates impressive capabilities, particularly in handling diverse and nuanced language patterns, considering factors such as training/prediction costs, processing time, and resource demands associated with running GPT-3.5, the rule-based approach emerges as an appealing choice for tasks involving SHI recognition and normalization. These findings highlight the importance of weighing the trade-offs between data-driven and knowledge-intensive approaches in selecting the most suitable methodology for specific tasks.

Table 8. Comparison between the ChatGPT and rule-based methods.

	GPT3.5 TURBO	Rule-based
Performance	0.775	0.877
Preparatory work	More expensive (The total cost including the experiment is NTD 1,953)	More labor intensive (Approximately 16 days, 5~9 h per day)
Match prediction time	Longer (40 min)	Shorter (30 s)
Cost	Expensive (OpenAI)	nil

5   Conclusions

The study investigates the case using two different distinctive approaches to solving the AICUP 2023 competition: LLMs based on the ChatGPT, and an alternative approach based on rules. The developed systems achieved impressive macro-F-scores of 0.752 and 0.866 for SHI recognition and 0.799 and 0.869 for temporal information normalization tasks, respectively. The ChatGPT method utilized the gpt-3.5-turbo-1106 model, chosen for its user-friendly implementation. After an initial training round, the model highlighted efficiency in addressing the two subtasks of the PS-EMR challenge, securing top rankings. We therefore further fine-tuned it with more training data to improve its ability to correctly recognize and normalize SHIs.

In contrast, the rule-based method demonstrated higher F-scores without the need for GPU-intensive model training. The method cost-effectively achieved competitive results, emphasizing the advantages of its lightweight architecture. Furthermore, the rule-based approach successfully addressed anomalies in the released dataset that the participants noticed, such as incorrectly labeled URL-related SHIs, shedding light on some potential errors in the competition's ground truths.

The study employed two methodologies to analyze the data, yet it is subject to certain limitations. The evaluation solely focused on their performance on a singular dataset, leaving uncertainty regarding their ability to apply to other datasets. The study did not examine potential biases that could sustain existing biases. In situations where rules are ambiguous or there is a lot of variation in the data, the rule-based approach may not be as efficient. Furthermore, the enhancements for both the ChatGPT and rule-based methods may involve augmenting the training data and patterns for rare SHI types and addressing challenges in LLM's output interpretation. These methods worked well together and showed how flexible it can be to use both advanced language models and rule-based approaches in the clinical de-identification task.

Acknowledgments. AI CUP 2023 Privacy Protection and Standardization of Electronic Medical Record competition (https://codalab.lisn.upsaclay.fr/competitions/15425) organizers provide the dataset. Our appreciation also goes to the University of New South Wales, Lowy Cancer Research Centre, and Health Science Alliance Biobank for granting access to the medical records utilized in this study. Additionally, we recognize the IW-DMRN workshop (https://www.sredhconsortium. org/sredh-competitions/sredhai-cup-2023/2024-iw-dmrn) and the SREDH Consortium (https:// www.sredhconsortium.org/) for their valuable contributions to our research endeavors.

Disclosure of Interests.   The authors have no competing interests.

References

1. Congress t. Health Insurance Portability and Accountability Act of 1996. ASPE – office of the assistant secretary for planning and evaluation, 20 August 1996
2. IW-DMRN.   https://www.sredhconsortium.org/sredh-competitions/sredhai-cup-2023/2024-iw-dmrn
3. Chen, A., Jonnagaddala, J., Nekkantti, C., Liaw, S.-T.: Generation of surrogates for de-identification of electronic health records. In: MEDINFO 2019: Health and Wellbeing e-Networks for All, pp. 70–73. IOS Press (2019)
4. Mir, T.H., et al.: Deidentification and temporal normalization of electronic health record notes using large language models: the SREDH/AI-Cup 2023 deidentification competition. In: Proceedings of the 2024 International Workshop on Deidentification of Electronic Medical Record Notes. Springer Nature, Kaohsiung (2024)
5.   Kedia, N., Sanjeev, S., Ong, J., Chhablani, J.: ChatGPT and beyond: an overview of the growing field of large language models and their use in ophthalmology. Eye (2024). https:// doi.org/10.1038/s41433-023-02915-z
6. Lee, Y.-Q., Chen, C.-T., Chen, C.-C., Lee, C.-H., Chen, P., Wu, C.-S., et al.: Unlocking the secrets behind advanced artificial intelligence language models in deidentifying chinese-english mixed clinical text: development and validation study. J. Med. Internet Res. **26**, e48443 (2024). https://doi.org/10.2196/48443
7. Liu, J., Gupta, S., Chen, A., Wang, C.K., Mishra, P., Dai, H.J., et al.: OpenDeID pipeline for unstructured electronic health record text notes based on rules and transformers: deidentification algorithm development and validation study. J. Med. Internet Res. **25**, e48145 (2023). https://doi.org/10.2196/48145
8. Jonnagaddala, J., Chen, A., Batongbacal, S., Nekkantti, C.: The OpenDeID corpus for patient de-identification. Sci. Rep. **11**(1), 19973 (2021). https://doi.org/10.1038/s41598-021-99554-9
9. Nlv, A., Aipeng, C., Batongbacal, S., Nekkantti, C., Dai, H.-J., Jonnagaddala, J.: Cohort selection for construction of a clinical natural language processing corpus. Comput. Methods Programs Biomed. Update **1**, 100024 (2021)
10. New embedding models and API updates. https://openai.com/blog/new-embedding-models-and-api-updates
11. Chang, N.W., Dai, H.J., Jonnagaddala, J., Chen, C.W., Tsai, R.T.H., Hsu, W.L.: A context-aware approach for progression tracking of medical concepts in electronic medical records. J. Biomed. Inform. **58**(S150–S157) (2015)
12. Wei, J., et al.: Fine-tuned language models are zero-shot learners. In: International Conference on Learning Representations, October 2021
13. Alkaissi, H., McFarlane, S.I.: Artificial hallucinations in ChatGPT: implications in scientific writing. Cureus **15**(2), e35179 (2023). https://doi.org/10.7759/cureus.35179

Advancing Sensitive Health Data Recognition and Normalization Through Large Language Model Driven Data Augmentation

Chia-Yi Chao$^{(\boxtimes)}$ [ID] and Cheng-Wei Lin [ID]

Intelligent System Laboratory, College of Electrical Engineering and Computer Science,
National Kaohsiung University of Science and Technology, Kaohsiung, Taiwan
joey5489joey@gmail.com

Abstract. Electronic Medical Record (EMR) text notes are a digital version of a patient's paper chart. It contains a comprehensive record of a patient's medical history. EMR text notes are designed to streamline healthcare processes, improve accuracy, and enhance patient care by providing easy access to up-to-date patient information for healthcare providers. The AI-cup 2023 competition for privacy protection and standardization of electronic medical records (EMR) has released a dataset annotated with sensitive health information (SHI) and temporal normalization values. This dataset aims to facilitate the development and evaluation of state-of-the-art natural language processing technologies for the task of privacy protection and standardization of EMR text notes. However, we observed that the annotation distribution for different SHI types is highly unbalanced. We, therefore, proposed a large language model (LLM)-powered data augmented approach to generate synthesized training instances to train an LLM based on the Pythia-410 m model released by EleutherAI. Combined with the pattern-based post-processing method, our team, TEAM_3917, achieved macro-F-scores of 0.8155 and 0.8065 for SHI recognition and temporal information normalization, respectively, which were officially ranked fourth during the competition.

Keywords: Natural language processing · de-identification · sensitive health information · temporal information normalization · data augmentation · pattern-based · large language model

1 Introduction

With the advancement of modern technology, healthcare systems increasingly rely on electronic medical records (EMR) text notes, which digitize all medical information. At this juncture, parsing these EMR text notes becomes crucial. Large Language Models (LLMs) have utilized various architectural models [1], bringing both opportunities and challenges in data privacy. While they have demonstrated impressive capabilities across various tasks, concerns about patient information privacy persist when using EMR text notes in the healthcare industry [2]. Therefore, it is important to adopt a method that is secure, highly accurate, and efficient to parse these data,

J. Jonnagaddala et al. (Eds.): IW-DMRN 2024, CCIS 2148, pp. 48–59, 2025.
https://doi.org/10.1007/978-981-97-7966-6_4

The AI CUP 2023 Privacy Protection and Electronic Medical Record Standardization (PS-EMR) competition [3–5] aims to address this issue by releasing de-identified and temporally standardized corpora. This competition tackles two tasks separately: de-identification and temporal information standardization. Participants need to annotate patients' Sensitive Health Information (SHI), along with their start and end positions, and normalize the time. The competition evaluates using precision, recall, and macro- and macro-F1-measure as the ranking metric [6]. We address this problem by proposing a data augmentation method based on large language models (LLMs) [7]. Additionally, we integrate pattern-based methods to handle more complex language phenomena and context-dependent corrections, achieving higher accuracy and fluency.

2   Methodology

2.1   Statistical Summary of the Released Dataset

The dataset released by PS-EMR text notes is OpenDeID v2 corpus comprises 3,244 pathology textual reports, consisting of 2,100 reports sourced from the OpenDeID v1 corpus, along with an additional 1,144 reports [8, 9], which were divided into training, validation, and test sets consisting of 1,734, 560, and 950 reports respectively. Each dataset contains annotations for sensitive health information (SHI) and temporary normalization information. Figure 1 summarizes the number of annotations in the released first and second-phase training set.

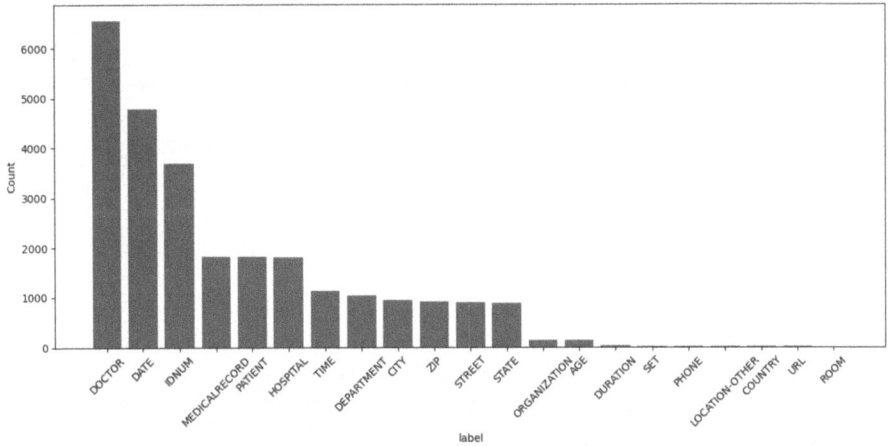

Fig. 1.   Statistics of released training set.

The annotations for different SHI types are highly unbalanced. Table 1 summarizes the most and least common annotations observed in the training set.

According to our experiments, we observed that the aforementioned imbalance issue significantly leads to the performance gap of the developed model especially in the recall rates. Inspired by the work presented by [10], Therefore, we proposed to employ

Table 1. Highest and Lowest Occurring SHI Annotations in the Training Set

Top 5 most common SHI types	Number	Top 5 least common SHI types	Number
DOCTOR	6541	ROOM	1
DATE	4777	URL	3
IDNUM	3677	COUNTRY	3
MEDICALRECORD	1824	LOCATIONOTHER	6
PATIENT	1819	PHONE	9

ChatGPT to generate synthetic examples to balance the training set. Specifically, we use prompts illustrated in Fig. 2 to direct ChatGPT to generate data with the same format but with different content.

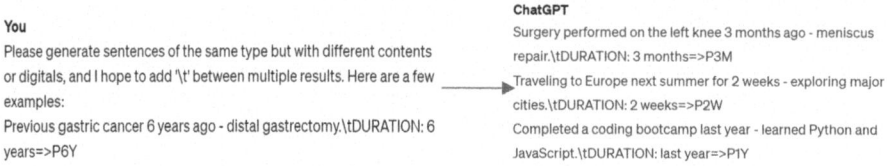

Fig. 2. An example of the prompt given to ChatGPT and the generated output.

With both the original data and data generated by ChatGPT, preprocessing of the dataset begins. To facilitate training and inference, each piece of data is converted into a file name, starting position, content, and answer. Additionally, we address the issue of data sentence segmentation. In the original data text, keywords are sometimes split across lines. Therefore, special attention is paid to such issues when converting the data into trainable data.

2.2   Training Strategy for Large Language Models

In our study, we fine-tuned the "EleutherAI/pythia-410m-deduped" model released by EleutherAI [11] on the released dataset. The model architecture is based on the Transformer [12], with the following key differences [13]. First, the rotary positional embeddings [14] were used to enhance the handling of sequence information, particularly in longer sequences. Afterward, parallel computation of attention and feed-forward can be computed in parallel and then summed. This parallel computation allows for the simultaneous processing of attention mechanisms and feed-forward propagation, as opposed to the sequential computation in traditional Transformers. The parallel computation can reduce the frequency of all-reduce operations, thereby improving the model's throughput. Finally, all dense layers are used to reduce implementation complexity.

To compile the dataset for fine-tuning, we preprocessed the training data by extracting the sentence along with the human-annotated annotations from the training set and format the training instance using the following format.

{START_SYMBOL}   sentence   {RESPONSE_INDICATOR}   annotations
{END_SYMOBL}.

We included the definitions for the STAR_SYMBOL, RESPONSE_INDICATOR, and END_SYMBOL in the tokenizer and treated them as special tokens to avoid further tokenization. Each sentence in the training set was therefore following the pre-designed format to form the training data for fine-tuning the target language model.

The optimization method we employed is AdamW [15] which is an enhancement of the Adam algorithm [16], which introduces a weight decay term to Adam, aiding in preventing model overfitting by decaying weights during gradient update steps. We utilized the AdamW optimizer provided by the PyTorch package, with a learning rate set to 0.00005. Additionally, we utilized a scheduler to adjust our learning rate, selecting the "linear warmup" algorithm. The algorithm begins with the learning rate starting from 0 and then linearly increasing to our preset learning rate. Subsequently, based on the number of steps specified when initializing this algorithm, the learning rate linearly decreases from our present value to 0. We chose this algorithm because our trained model is relatively large, and the output of this task was also scattered, potentially leading to higher losses, making it challenging to find the right direction for convergence. Therefore, we aimed to start with a smaller learning rate initially to facilitate gradual learning of the model, leading it toward the optimal loss. Then, gradually decreasing the learning rate from a higher value aids the model in final convergence. The number of the warmup steps was set to one-tenth of our total steps, and the number of the training steps was set to our total steps in our experiments.

2.3 Pattern-Based Post-processing

During the generation process of the fine-tuned models, there is a possibility that the model inevitably produces some "dirty" data, such as duplicate outputs, lengthy gibberish, or incomplete outputs. Through post-processing methods, we cleaned up some dirty data before outputting the results. Figure 3, second item, demonstrates the effectiveness of this general post-processing approach.

Furthermore, based on our analysis of the augmented-training results, we found that regardless of training efforts, the overall recall rate remained relatively low for the less common SHI types shown in Table 1. Therefore, we developed a pattern-based post-processing method to recognize the missing SHIs and integrate the results with the language model's output to improve the overall recall rates.

We developed pattern-based methods for SHI types including COUNTRY, URL, DURATION, SET, LOCATION-OTHER, PHONE, and AGE. These are predominantly what we refer to as unpopular labels. This is primarily due to their inherently lower scores and their infrequent occurrence. Although our experiment results show that including this method potentially reduces the precision rate, the decrease is relatively minor.

3 Results

Our team's workflow in the competition includes five steps: data preprocessing, data augmentation, training the model, data post-processing, and a pattern-based approach. Figure 3 illustrates this workflow.

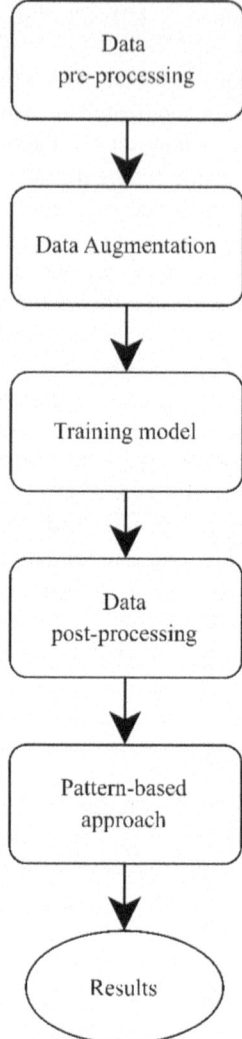

Fig. 3. Workflow diagram

Figure 4 illustrates the effectiveness of the proposed methods. The baseline configuration relies only on fine-tuning processing of the original published dataset, training 13 times using a 410M size model without applying any post-processing. As one can see the overall macro-F-score is significantly low and the baseline score is only 0.6553. For the post-processing configuration, we employed a post-processing method to filter out noisy output. The score was then improved to 0.6961. In the LLM-based augmentation configuration, we show the performance of the same 410M model fine-tuned on the augmented dataset. It can be observed that training with the synthesized data augmented for the less common SHI types can significantly improve the F-score by 0.20. Finally, we include the pattern-based post-processing method to further involve the F-score from 0.85 to 0.921. The results demonstrate that even though the LLM-powered augmentation method can improve the model's F-score, the inclusion of the pattern-based post-processing can further improve the performance.

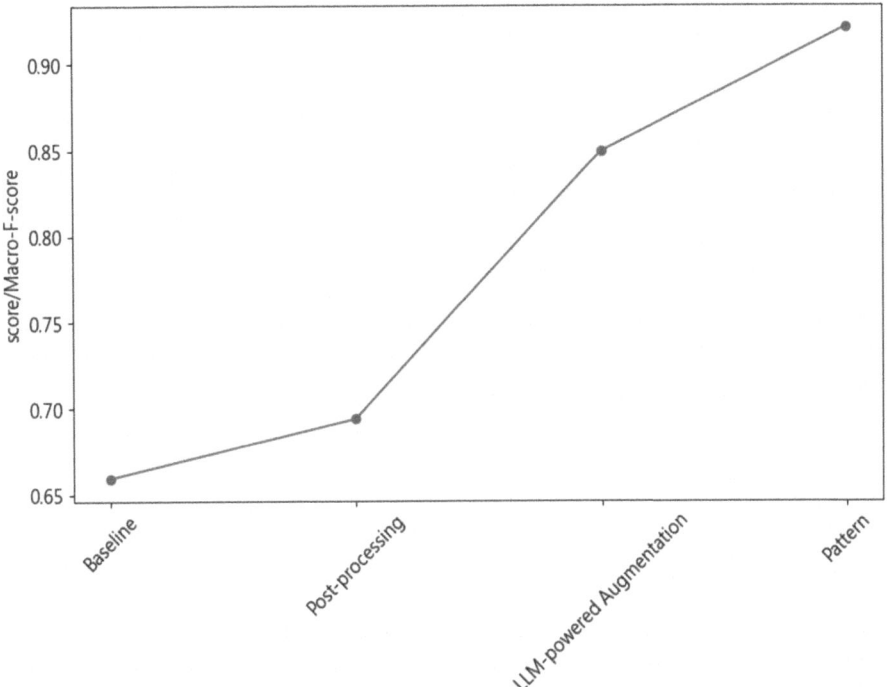

Fig. 4. The differences in scores at various stages of validation data.

Through the method explained above, we have achieved comparative results. For subtask 1 for SHI recognition, a macro-F-score of 0.8155 was obtained, and in subtask 2 for temporal information normalization, the proposed method achieved a macro-F-score of 0.8065. Table 2 shows the detailed micro-F-scores for each SHI type.

Table 2. SHI recognition and temporal information normalization F-scores on test data

Subtask 1	Micro-F-Score	Subtask 2	Micro-F-Score
IDNUM	0.9710	DATE	0.7968
MEDICALRECORD	0.8532	TIME	0.6206
PATIENT	0.9394	DURATION	0.9090

(continued)

Table 2. (*continued*)

Subtask 1	Micro-F-Score	Subtask 2	Micro-F-Score
STREET	0.9597	SET	0.8889
CITY	0.9450		
ZIP	0.8737		
DATE	0.9347		
TIME	0.7776		
DEPARTMENT	0.8480		
HOSPITAL	0.8794		
DOCTOR	0.8923		
STATE	0.9746		
DURATION	0.8695		
ORGANIZATION	0.5903		
SET	0.8889		
AGE	0.8602		
URL	0.9400		
LOCATIONOTHER	0.5000		
PHONE	0.0000		

4   Discussions

4.1   Data Augmentation

Through observing the data distribution and considering that the scoring criterion for this competition is the average score across all types, our strategy is to generate more cases for formats where the model performs poorly and fewer cases for formats where the model performs well. This ensures that the model learns a more balanced data distribution during training, preventing a significant drop in the overall score due to specific format deficiencies. During the competition, we made data argumentation for 11 types, including DOCTOR, DATE, IDNUM, TIME, ZIP, AGE, ORGANIZATION, DURATION, SET, PHONE, LOCATION-OTHER. Table 3 illustrates the contrast in the quantities of each label before and after augmentation.

Table 3. Data for each SHI

SHI types	Original number	Generated number	Increase rate	Popular or Unpopular
DOCTOR	8400	100	0.0119	Popular
DATE	6300	100	0.0158	Popular
IDNUM	4850	100	0.02	Popular

(*continued*)

Table 3. (*continued*)

SHI types	Original number	Generated number	Increase rate	Popular or Unpopular
TIME	1650	260	0.157	Popular
ZIP	1100	55	0.0283	Popular
AGE	190	620	3.263	Unpopular
ORGANIZATION	130	650	5.0	Unpopular
DURATION	50	715	14.3	Unpopular
SET	28	359	12.821	Unpopular
PHONE	14	520	5.349	Unpopular
LOCATION-OTHER	8	735	2.598	Unpopular

4.2 Performance Comparison Between Different Model Sizes

EleutherAI provides a range of Pythia models with different parameter sizes ranging from 70M, 160M, 410M, 1B, 1.4B, 2.8B, 6.9B, and 12B [17]. In this study, we primarily trained models of sizes 70M, 160M, and 410M. We trained the three models with the same expanded training data, methods, and hyper-parameter settings. Figure 5 illustrates the performance comparison among the three models of different sizes, showing that larger models achieve better performance. Therefore, we selected the pre-trained model with 410M parameters in the following experiments.

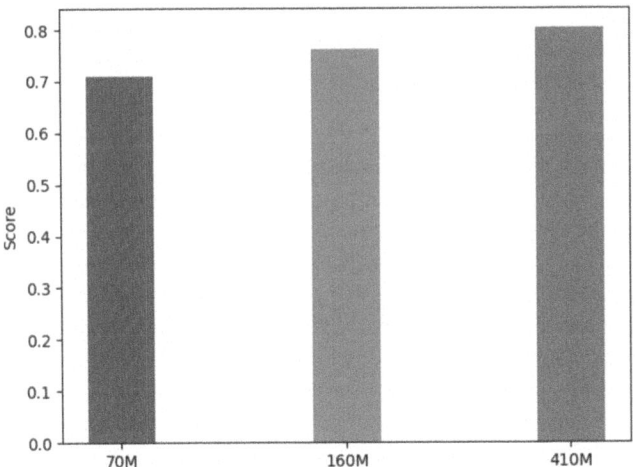

Fig. 5. Scores for models of different sizes.

Figure 6 illustrates the results of our experiments on the 410M model trained with three different algorithms: SGD, Adam, and AdamW. Each algorithm was trained 13 epochs with the same original training dataset. From the results, it can be seen that compared to SGD and Adam, AdamW's optimization strategy leads to better convergence. In this competition, the AdamW algorithm performed best, so it was selected in the final algorithm strategy.

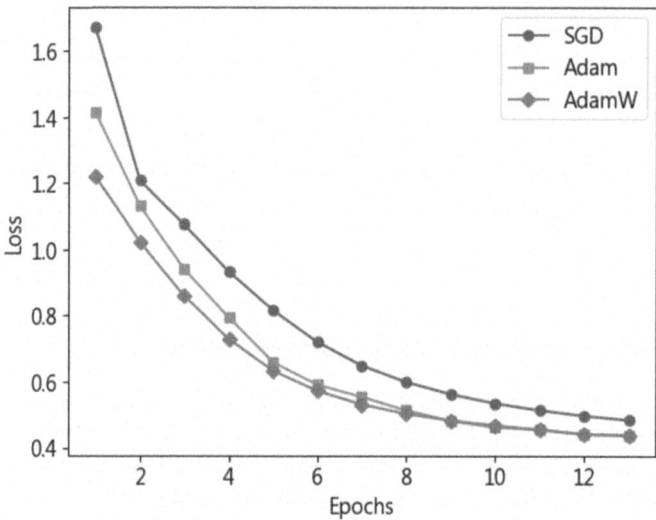

Fig. 6. Validation loss comparison between different training algorithms for the 410M pre-trained model.

4.3 Effectiveness of the Proposed LLM-Powered Data Augmentation and Post-processing Methods

During the challenge, we focused on addressing the imbalanced issue for less common SHI types that appeared in the released dataset and the model trained with the augmented dataset indeed outperformed the model trained with the original training set. However, as we analyzed the outputs of the developed models and the statistics of the test set, we observed that there is a potential for further boosting the performance of our system by improving the pattern-based approach. In the competition, as mentioned earlier, we only applied a pattern-based approach to seven SHI types. After the competition, we extended our pattern-based approach to include other unpopular types and even some popular SHI types. We found that popular labels still exhibit extremely high repeatability, allowing us to apply pattern-based methods similar to what we did for unpopular labels. However, dealing with such large volumes of data not only requires a significant amount of time to search for every possible format but also makes it easier to overlook certain formats compared to unpopular labels. This can lead to a decrease in precision for that label. Despite these drawbacks, by applying pattern-based methods to labels such as IDNUM, MEDICALRECORD, CITY, and HOSPITAL, we managed to improve our score from the initial 0.815 to 0.847.

Both our study and [18] indicate that large language models based on the Transformer architecture are effective in identifying and normalizing SHI. The LLM-powered data augmentation method proposed by us can address the shortcomings of the original database, preventing the model from overfitting to a large volume of frequent data and thus ignoring labels with lower occurrence frequencies. While pattern-based methods may consume significant time, they serve to further extend the limits of LLMs. We have reason to believe that perhaps it is unnecessary to push one method to its limits for handling such SHI recognition cases, but rather, the fusion of LLMs and pattern-based methods may be a future trend.

The integration of LLMs and pattern-based post-processing techniques holds great potential for improving the recognition and normalization of SHI. This combination offers a promising avenue for enhancing NLP model performance in SHI-related tasks, with extensive applications in healthcare data processing and privacy protection. Further research and optimization in this area are crucial for advancing the capabilities of SHI recognition and normalization systems, ultimately contributing to more accurate and efficient healthcare data management.

5 Conclusions

We proposed an LLM-powered data augmentation approach to improve the performance of SHI recognition and normalization. We addressed the challenge of highly unbalanced annotation distributions for different SHI types in EMR text notes by leveraging LLMs for data augmentation. Our approach enhances model robustness and prevents overfitting to frequent data while neglecting less common SHI labels.

The results highlight the effectiveness of LLMs in handling SHI recognition and normalization tasks. Furthermore, the integration of pattern-based post-processing techniques further enhances model performance. We anticipate that future research will continue to explore the fusion of LLMs and pattern-based methods, providing a promising avenue for improving NLP model performance in healthcare data processing and privacy protection.

In conclusion, the integration of LLMs and pattern-based post-processing techniques offers a robust approach to enhancing the recognition and normalization of SHI. Further research and optimization in this area are essential for advancing the capabilities of healthcare data management systems, contributing to improved patient care and privacy protection.

Acknowledgments. Our team would like to thank the organizers of the AI CUP 2023 Autumn Competition (https://codalab.lisn.upsaclay.fr/competitions/15425), Additionally, we recognize the IW-DMRN workshop (https://www.sredhconsortium.org/sredh-competitions/sredhai-cup-2023/2024-iw-dmrn) and the SREDH Consortium (https://www.sredhconsortium.org/) for their valuable contributions to our research endeavors.

Disclosure of Interests. The authors have no competing interests.

References

1. Reuben, A.S., Jarrel, C.Y.S., Ke, C., Lincoln, L., Wei, L., Justin, Y.: Generative Large Language Models for Detection of Speech Recognition Errors in Radiology Reports Jan 24 (2024)
2. Zou, S., He, J.: Large language models in healthcare: a review. In: 2023 7th International Symposium on Computer Science and Intelligent Control (ISCSIC) (2023)
3. AICup [Internet]. https://www.aicup.tw/copy-of-ai-cup-2023-1
4. SREDH Consortium [Internet]. https://www.sredhconsortium.org/
5. IW-DMRN [Internet]. https://www.sredhconsortium.org/sredh-competitions/sredhai-cup-2023/2024-iw-dmrn
6. Codalab [Internet]. https://codalab.lisn.upsaclay.fr/competitions/15425
7. Whitehouse, C., Choudhury, M., Aji, A.F.: LLM-powered data augmentation for enhanced crosslingual performance. arXiv preprint arXiv:2305.14288 (2023)
8. Jitendra, J., Aipeng, C., Sean, B., Nekkantti, C.: The OpenDeID corpus for patient de-identification, 7 October 2021
9. Naga, L.V., Alla, A.C., Sean, B., Chandini, N., Hong-Jie, D., Jitendra J.: Cohort selection for construction of a clinical natural language processing corpus 2021 [Internet]. https://doi.org/10.1016/j.CMPBUP.2021.100024
10. Ruixiang, T., Xiaotian, H., Xiaoqian, J., Xia, H.: Does synthetic data generation of LLMs help clinical text mining?, 10 April 2023
11. Biderman, S., et al.: Pythia: a suite for analyzing large language models across training and scaling. In: International Conference on Machine Learning, pp. 2397–2430. PMLR, July 2023
12. Ashish, V., et al.: Attention Is All You Need, 12 June 2017
13. Sid, B., et al.: GPT-NeoX-20B: an open-source autoregressive language model
14. Su, J., Ahmed, M., Lu, Y., Pan, S., Bo, W., Liu, Y.: RoFormer: enhanced transformer with rotary position embedding. Neurocomputing **568**, 127063 (2024)
15. Loshchilov, I., Hutter, F.: Decoupled weight decay regularization. In: International Conference on Learning Representations, September 2018
16. Kingma, D.P., Ba, J.: Adam: a method for stochastic optimization. arXiv preprint arXiv:1412.6980 (2014)
17. EleutherAI Pythia scaling suite [Internet]. https://huggingface.co/collections/EleutherAI/pythia-scaling-suite-64fb5dfa8c21ebb3db7ad2e1
18. Jiaxing, L., et al.: OpenDeID Pipeline for Unstructured Electronic Health Record Text Notes Based on Rules, and Transformers: Deidentification Algorithm Development and Validation Study, 6 December 2023 [Internet]. https://www.jmir.org/2023/1/e48145
19. Mir, T.H., et al.: Deidentification and temporal normalization of electronic health record notes using large language models: the SREDH/AI-Cup 2023 deidentification competition. In: 2024 International Workshop on Deidentification of Electronic Medical Record Notes (2024). Kaohsiung, Taiwan: Springer Nature
20. Chen, A., Jonnagaddala, J., Nekkantti, C., Liaw, S.-T.: Generation of surrogates for de-identification of electronic health records. In: MEDINFO 2019: Health and Wellbeing e-Networks for All, p. 70-3. IOS Press (2019)
21. 2024 International workshop on deidentification of electronic medical record notes (IW-DMRN) [Internet]. https://www.sredhconsortium.org/sredh-competitions/sredhai-cup-2023/2024-iw-dmrn
22. Moharasan, G., Ho, T.B.: Extraction of temporal information from clinical narratives. J. Healthc. Inform. Res. **3**(2), 220–244 (2019)
23. Silvestri, S., Esposito, A., Gargiulo, F., Sicuranza, M., Ciampi, M., De Pietro, G.: A big data architecture for the extraction and analysis of EHR data (2019). https://ieeexplore.ie

24. Li, L., Jin, L., Jiang, Z., Song, D., Huang, A.D.: Biomedical named entity recognition based on extended recurrent neural networks. In: 2015 IEEE International Conference on Bioinformatics and Biomedicine (BIBM), Washington, DC, USA (2015)

25. Morwal, S., Jahan, N., Chopra, D.: Named entity recognition using Hidden Markov Model (HMM). Int. J. Nat. Lang. Comput. (IJNLC) 1(4) (2012). https://papers.ssrn.com/sol3/papers

26. Viani, N., et al.: Temporal information extraction from mental health records to identify duration of untreated psychosis. J. Biomed. Semant. 10(2) (2020)

27. Kajiyama, K., Horiguchi, H., Okumura, T., Morita, M., Kano, Y.: De-identifying free text of Japanese electronic health records. J. Biomed. Semant. 11(1), 1–12 (2020)

28. Uzuner, O., Luo, Y., Szolovits, P.: Evaluating the state-of-the-art in automatic de-identification. J. Am. Med. Inform. Assoc. 14(5), 550–563 (2007)

Privacy Protection and Standardization of Electronic Medical Records Using Large Language Model

Chao-Long Huang ⓘ, Babam Rianto ⓘ, Jun-Teng Sun ⓘ, Zheng-Xin Fu ⓘ, and Chung-Hong Lee⁽✉⁾ ⓘ

National Kaohsiung University of Science and Technology, Kaohsiung 807618, Taiwan
`leechung@mail.ee.nkust.edu.tw`

Abstract. Recently, the widespread application of electronic medical records (EMRs) has made protecting patients' personal privacy information crucial and highly important. However, the sources of these EMRs are different, resulting in various time formats, and this inconsistency makes historical data challenging to read. Therefore, the key to solving the problem lies in two major issues in medical information: de-identification of sensitive health information (SHI) as defined by the Health Insurance Portability and Accountability Act (HIPAA) and normalization of time according to the ISO 8601 standard. To identify the items in case reports, we used the powerful capabilities of large language models (LLMs) as the foundation. We developed a set of algorithms based on In Context Learning (ICL), which ensures that the model inference process is always in the correct order while allowing the features contained in invalid labels to be utilized. The efficacy of our method has been validated through its exceptional performance in a real-world competition, attaining an macro-F-measure of 93.51% on the de-identification task and remarkably achieving over 92% accuracy for time normalization.

Keywords: De-identification · Time Normalization · Large Language Models · Electronic Medical Records · Sensitive Health Information

1 Introduction

In recent years, artificial intelligence (AI) has gained significant interest in healthcare applications. This trend is exemplified by the AI-Cup competition series, a collaborative effort between the Taiwanese government, institutions, and top universities, alongside the Secure Research Environment for Digital Health (SREDH) Consortium [1]. The 2024 International Workshop on De-identification of Electronic Medical Record Notes (IW-DMRN) [2] focused on tasks like retrieving patient privacy information and normalizing time data. Held at the Department of Electrical Engineering, National Kaohsiung University of Science and Technology (NKUST) in Taiwan [3], the workshop saw our team, AUSTIN 2526 (TEAM_3951), achieve second place in task 1 and fourth place in task 2, ranking us second overall out of 103 teams [4]. This competition challenged participants to develop and apply AI technologies using de-identified EMR text notes data from real medical institutions [5].

© The Author(s), under exclusive license to Springer Nature Singapore Pte Ltd. 2025
J. Jonnagaddala et al. (Eds.): IW-DMRN 2024, CCIS 2148, pp. 60–71, 2025.
https://doi.org/10.1007/978-981-97-7966-6_5

In the medical field, EMRs store a patient's clinical history, demographics, allergies, family history, and other relevant information. This data helps doctors understand a patient's medication history, past treatments, and referrals, leading to safer and more effective care plans. EMRs also facilitate diagnoses and treatment planning. For example, Alla et al. [6] explored the construction of a deidentification corpus using pathology reports for cancer patients. In their study, 2,100 pathology reports were extracted from the electronic health records (EHRs) of 1,833 patients (518 male and 1,313 female) using the Health Level-7 (HL7) messaging standard. They reported that the most represented age group was 60–70 years, with 872 patients. Their analysis highlights significant challenges in deciphering segment information from HL7 messages collected across various hospitals, compounded by the inconsistent tagging within the HL7 messages themselves. This inconsistency often rendered the identification of reports that met pre-set criteria difficult. Furthermore, the healthcare system is increasingly adopting AI technologies, particularly LLMs. Nori et al. [7] conducted a comprehensive evaluation of GPT-4, assessing its applicability and efficacy in medical competency examinations and benchmark datasets. Despite being a general-purpose LLM without specialized training for medical domains, GPT-4 demonstrated remarkable proficiency. The evaluation included analyses using official practice materials for the United States Medical Licensing Examination (USMLE), which is a critical three-step examination program for clinical competency and licensure in the U.S. Additionally, the study extended to the MultiMedQA benchmark datasets to further gauge performance across diverse medical queries. Wang et al. [8] introduced a novel framework for the integration of LLMs into medical-image computer-aided diagnosis (CAD) networks. This methodology capitalizes on the unique capabilities of LLMs, such as advanced medical domain knowledge and logical reasoning, to enhance the performance of various CAD network functionalities, including diagnosis, lesion segmentation, and report generation. By synthesizing and reformatting the output of these networks into natural language text, the proposed framework aims to bridge the gap between complex medical imaging data and clinical application, making the resulting information more accessible and comprehensible for patients. Wang et al. [9] developed a novel system called LLMs Augmented with Medical Textbooks (LLM-AMT), which is designed to boost the proficiency of LLMs in specialized medical domains. This system incorporates a set of plug-and-play modules including a Query Augmenter, a Hybrid Textbook Retriever, and a Knowledge Self-Refiner, which collectively embed authoritative medical knowledge directly into the LLM framework. Additionally, the accuracy gains ranged from 11.6% to 16.6%. Remarkably, when using GPT-4-Turbo as the base model, LLM-AMT surpassed the performance of the specialized Med-PaLM 2 model, which had been pre-trained on a substantial medical corpus, by 2–3%. This finding underscores the effectiveness of using authoritative medical textbooks as a knowledge base, which, despite being 100× smaller in size compared to a massive dataset like Wikipedia, provided a 7.8%–13.7% boost in performance within the medical domain. Jin et al. [10] proposed a Health-LLM for predicting potential diseases and offering personalized health advice. However, they emphasize that LLMs are not replacements for human expertise, but rather powerful tools that can improve efficiency and accuracy in various tasks. EMRs play a crucial role in this context. By extracting features from EMR text notes, we can predict patient disease risk, allowing doctors to take preventive measures. This personalized approach can improve treatment outcomes and reduce healthcare costs. However, AI technologies often involve personal data, raising privacy concerns [11–13]. To prevent privacy breaches, the Health Insurance Portability and Accountability Act (HIPAA) sets strict standards for protecting patient data [14]. HIPAA regulations mandate that healthcare organizations safeguard patients' SHI and prevent unauthorized access or use. Manual de-identification is time-consuming, prone to errors, and might leave some SHI behind, increasing the risk of patient identification [14]. While previous research has identified limitations in many de-identification methods, statistical learning-based systems offer distinct advantages for de-identifying free text. These systems rely on labeled training data to detect patterns and identify and remove protected health information from raw data [15]. However, the growing volume of digitized data makes ensuring HIPAA compliance more complex, rendering manual de-identification methods inefficient and hindering technological advancements. Furthermore, the literature has shown that deep learning techniques have significantly improved the detection of sensitive health information within data containing personal identity information. Yang et al. [16] proposed a deep learning model called LSTM-CRFs. This model combines Conditional Random Fields (CRFs) and integrates FastText technology, achieving effective de-identification in both general and clinical domains. These methods enhance the reliability of Named Entity Recognition (NER) in the biomedical field in several ways. By integrating CRFs, de-identification models can comprehensively understand and process sensitive information data, leading to improved accuracy and better handling of complex scenarios [17, 18]. The rapid rise of LLMs has also opened new avenues for de-identification technology [19]. Language generation models, such as GPT, have become adept at understanding and generating text due to their pre-training on massive amounts of language data [20]. Furthermore, Few-Shot Learning (FSL) has gained popularity, allowing language models to learn tasks from just a few demonstration examples [21]. Dong et al. [21] delve into how FSL focuses on the distribution of training data, learning mechanisms, error bounds, and the influence of prior feature biases. They highlight the similarities between FSL and gradient descent, as well as the functional components within transformers that facilitate FSL, ultimately enhancing its performance. Hu et al. [22] evaluated ChatGPT's ability to perform zero-shot learning for clinical NER using the 2010 i2b2 challenge dataset. While their study acknowledges that ChatGPT's zero-shot performance falls short of a fine-tuned BioClinicalBERT-based clinical NER model, it still achieves a satisfactory level (with an F1 score of 0.628 in relaxed matching). Shyr et al. [23] explored using zero-shot and few-shot techniques in NER tasks involving rare disease entities. ChatGPT obtained F1 scores of 0.472 and 0.407 in the zero-shot scenario using simple sentence and structured list prompts, respectively. The performance improved in the few-shot scenario, yielding F1 scores of 0.591 and 0.469. Further improvement was achieved by selecting training text based on similarity scores, resulting in F1 scores of 0.610 and 0.544. LLM models are being applied to more complex text analysis and data understanding tasks in de-identification. By combining them with large language models, researchers are exploring ways to improve the efficiency of context understanding and information processing within the de-identification process [24]. This integration makes the models more adaptable and able to handle diverse data formats and contexts better, further enhancing the flexibility of de-identification technology in practical applications [25].

Despite the numerous research achievements of LLMs in the medical field, their application research in medical de-identification remains relatively limited. Therefore, the primary purpose of this study is to promote the exploration of using large language models for de-identification and to improve the efficiency and accuracy of medical data management in this way. In this work, we utilize the LLM model and combine it with the in-context learning method through the sliding window technique. This approach establishes a new way for handling SHI text note data, potentially surpassing existing methods. The development of this technology is likely to trigger more related research and industrial applications, further driving progress in medical research and health information management.

2 Methodology

2.1 Dataset

In this study, the EMRs data source was provided by the Health Science Alliance Biobank at the Lowy Cancer Research Centre, University of New South Wales (UNSW), Australia. The OpenDeID v2 dataset comprises a detailed set of 3,244 pathology reports 2100 reports from OpenDeID v1 corpus+1144 additional reports [26]. In this dataset, each EMR text notes file contains a substantial amount of text notes [4]. Inputting the entire EMRs directly into the model can impair the effectiveness of the model's attention mechanism, potentially leading to diminished results. Consequently, it is imperative to thoughtfully approach text segmentation. This strategy enables the system to reason more effectively and accurately identify the labels requiring de-identification within the EMR text notes. Notably, the positioning of labels within these datasets is relatively uniform (see Fig. 1).

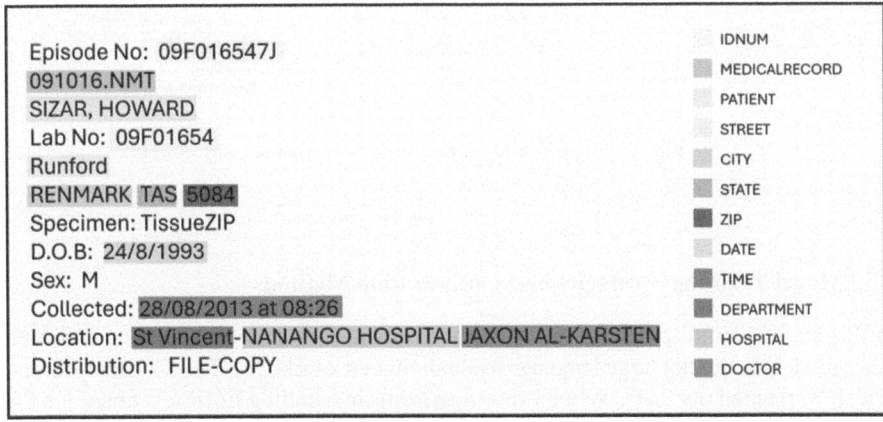

Fig. 1. An EMR data sample.

2.2 Combining in Context Learning Technique with Sliding Window Algorithm

In LLMs, combining context is a very important technique, which can give the model broader information for reasoning and achieve better results. In this work, we developed an algorithm that combines in-context learning with the Sliding Windows method [21]. The objective is to enhance the correlation between sequences and establish a connection between them and their respective labels. Simultaneously, the sliding windows method is employed to comprehensively understand the start and end of each sequence, enabling the model to base its inferences consistently on the preceding text. This approach allows each line of text to be learned twice. However, it could lead to the model learning many irrelevant labels. To mitigate this, we introduced a parameter to eliminate these irrelevant labels. Our method offers two distinct advantages for large language models. First, it utilizes the information interconnected among rows to integrate the substantial number of irrelevant labels in the dataset into the sequence of relevant labels, providing more meaningful information to these relevant labels. Second, it enables the model to concentrate on pertinent labels. Approximately 75,000 lines of data can be broken down in the dataset, of which around 54,000 are considered irrelevant labels. In large language models, if a small amount of data is provided, the model may incorrectly categorize it as irrelevant. This is because irrelevant labels are typically easier to learn than logically complete labels. Nevertheless, an excess of irrelevant labels can impede the model's learning of logically complete labels, diminishing its overall performance. Our training method uses the overlapping parts in Windows for validation, as depicted in Fig. 2. This technique effectively assesses the model's ability to infer implicit messages and helps prevent overfitting.

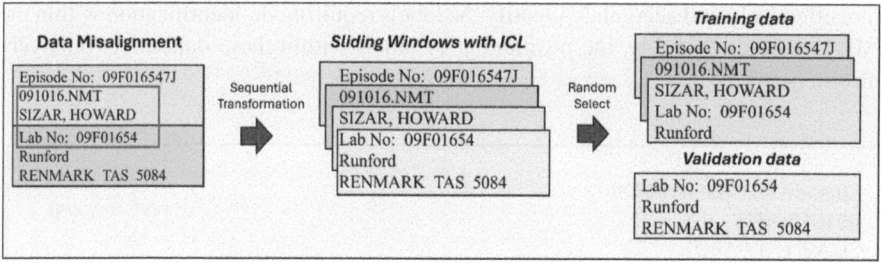

Fig. 2. Preprocessing for EMR data segmentation.

2.3 Model Training Strategies and Optimization Methods

In this study, we utilized the QWEN 7B and 14B models [27], as they outperformed other models in crucial large language evaluation metrics like MMLU. Currently, Wang et al. [28] pointed out that QWEN-14B outperforms in handling EMR text notes. On the other hand, QWEN-7B significantly outperforms BERT and BGE-M3 in all metrics for handling the embedding clinical notes [29]. We have implemented the Fully Sharded Data Parallel (FSDP) technique to enhance the stability of the model. In this study, we propose two measures to counteract the issue of overfitting [30]. Firstly, we employ the NEFTune method to introduce uniformly distributed noise, applying this noise to the model's embedding layer [31]. This approach prevents the model from overly memorizing the specific details and noise within the training set, thus enabling it to adapt more effectively to new data. The implementation of this technique also reduces the model's sensitivity to particular inputs and prevents it from developing overly complex representations at an early stage. Another strategy involves randomly extracting windows and utilizing them as a basis for validating the optimal model performance.

In the inference phase of the model, we consider selecting a greedy decoding strategy the optimal choice. This is because the output probability of greedy decoding is relatively stable, which is crucial for ensuring the stability and reliability of the model. Compared to other inference strategies, using a greedy decoding strategy does not significantly degrade the model's effectiveness while maintaining overall performance stability.

In addition to the greedy decoding strategy, we incorporate an additional layer of rule-based post-processing model after the inference phase of the model. Since the original model occasionally makes erroneous predictions for some labels that should be straightforward to classify, such issues can lead to significant deviations in the final results, thereby impacting the overall performance and accuracy of the model. To overcome this challenge, we have introduced several processing steps to ensure the correct identification of labels and effective de-identification. The purpose of this additional post-processing layer is to ensure that easily identifiable known labels are not misclassified, thereby further enhancing the accuracy and reliability of the model. Through this reinforcement step, we are better equipped to handle misclassification issues in specific scenarios, thereby improving the model's overall performance. We have focused on improving the following categories of labels.

During the time normalization process, we often face challenges related to the overlap between data and time labels. This can result in instances where time is mistakenly identified as data during model inference. In Fig. 3, a rule-based model correction method is described. It illustrates that to address this issue, it is necessary to carefully examine the preceding and succeeding text segments when classifying a label as data, looking for the presence of time formats to ensure accurate identification and effective control of PHI label application. In the de-identification task, similar issues arise, such as the overlap between medical records and ID number labels. However, due to the presence of XXX suffixes in medical record labels, we can relatively easily manage these overlaps.

The performance evaluation of the proposed model involves several metrics, namely precision, recall, and F1-measure. The metrics consisting of several variables which is True Positives (TP): The number of positive instances correctly predicted as positive, False Positives (FP): The number of negative instances incorrectly predicted as positive and (False Negatives (FN): The number of positive instances incorrectly predicted as negative. The F1 measure is the harmonic means of precision and recall, providing a balance between them. It is especially useful when the class distribution is unbalanced. These indicators provide a comprehensive assessment of the system's effectiveness.

$$Precision = \frac{TP}{TP + FP} \tag{1}$$

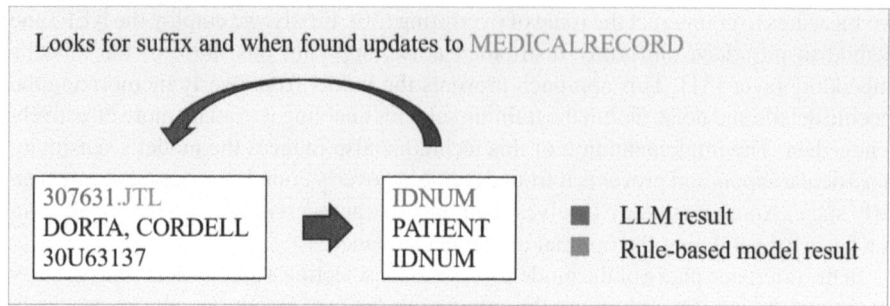

Fig. 3. Rule-based model Correction method.

$$Recall = \frac{TP}{TP + FN} \tag{2}$$

$$F1-measure = 2 \times \frac{Precision \times Recall}{Precision + Recall} \tag{3}$$

3 Result

In this experiment, the proposed approach demonstrated outstanding performance in the de-identification task, facilitating the extraction of a substantial amount of SHI from electronic reports. The results, as depicted in Table 1, show that our technology has reached a high level of proficiency. Evaluation during the competition was carried out using macro-F-measure, a metric that evaluates the risk of patient information leakage, thereby effectively validating the model's capabilities. The competition results confirm that our research has reached a professional standard, demonstrating its potential for real-world applications.

In this work, we employed the Qwen-14B model for the time normalization task. As indicated in Table 1, the Recall value for the "LOCATION-OTHER" category was not particularly high. This outcome was not a consequence of flaws in our strategy design but rather due to the presence of answers in the test set that were not previously exposed in the training and validation sets. Nonetheless, our model was able to make judgments based on sequence positioning and successfully identified 25% of these data. The performance on the other indicators was uniformly excellent. For the initial phase of time normalization in the competition, we used the Qwen-7B model for experimentation. However, the model's relatively limited parameters made it challenging to interpret the prompts we used. Consequently, we conducted more thorough experiments and upgraded to the Qwen-14B model. We retained our previous identification strategy during this stage. In addition, we integrated the Chain-of-Thought (COT)-few-shot method into the Prompting approach [32]. This method can guide the model to perform more effective reasoning. Specifically, we guide the model to the positions where a small number of time labels will appear so that it can enhance its ability to identify these labels. This strategy was mainly aimed at addressing the small number of labels in the DURATION and SET categories. As shown in Table 2, both labels achieved favorable results.

Table 1. Qwen de-identification result

Label	Precision	Recall	F-measure	Support
PATIENT	0.9916	0.9930	0.9923	716
DOCTOR	0.9784	0.9567	0.9674	3327
DEPARTMENT	0.9368	0.9212	0.9290	419
HOSPITAL	0.9094	0.9641	0.9359	1198
ORGANIZATION	0.9859	0.9459	0.9655	74
STREET	0.9913	0.9970	0.9942	334
CITY	0.9918	0.9839	0.9878	373
STATE	0.9969	0.9969	0.9969	332
ZIP	0.9971	0.9971	0.9971	353
LOCATION-OTHER	1	0.5	0.6667	6
AGE	0.9607	0.9607	0.9607	51
PHONE	1	1	1	1
URL	1	1	1	4
MEDICALRECORD	0.7879	1	0.8814	747
IDNUM	0.9835	0.9886	0.9861	2120

Table 2. Qwen time-normalization result

Label	Precision	Recall	F-measure	Support
DATA	0.9571	0.9796	0.9682	2459
TIME	0.8858	0.8198	0.8515	470
DURATON	1	1	1	12
SET	1	0.8	0.8889	5

After normalizing the time, the final results did not meet our expectations, especially in standardizing the "TIME" aspect. This can be attributed to the ambiguity in the dataset, which necessitates providing sufficient contextual cues to determine the originating unit of this EMR text notes and explicitly indicate whether the time is in the morning or afternoon. Additionally, variations in the order of months and dates significantly increased the difficulty of review. Despite manual inspection of these labels, identifying specific patterns remained quite challenging. Further analysis of the "TIME" labels revealed that the ratio of MM/DD/YYYY to DD/MM/YYYY date formats was approximately 3:7. This discrepancy in formats added additional complexity to the model's inference process, requiring the utilization of larger window sizes to adequately comprehend the characteristics of these labels. Through this approach, we ultimately succeeded in normalizing the time in these highly similar EMR data, achieving a macro-F-measure of 92%.

4 Discussion

In our research, we not only focus on improving the accuracy of the model but also on enhancing the reasoning speed to ensure a proportional increase alongside accuracy. While hardware upgrades offer a solution, they are not always feasible, especially in resource-constrained regions or scenarios. Therefore, there is a pressing need to optimize model performance within limited hardware constraints, facilitating the widespread adoption and application of the technology.

There are alternative approaches beyond NEFTune technology that can achieve similar effects. For example, one could adjust the learning rate differently for each neural network layer to fine-tune its performance. However, this approach might require significant time to search for the most suitable hyperparameters, especially for larger language models with more layers, making the tuning process more challenging. Our NEFTune method can efficiently enable the model to self-adapt and find better solutions. Additionally, we observed suboptimal performance of COT few-shot on the QWEN 7B model. This necessitates stronger reasoning capabilities, which the 7B model lacks on the reasoning [33]. To address this issue, we plan to expand the model to QWEN 14B, aiming for better adaptation to such reasoning methods, thereby enhancing overall performance to meet real-world application demands. This initiative will aid in more effectively leveraging this technology across different domains and application scenarios, thereby advancing and broadening its utilization in relevant fields.

Notably, de-identification of EMR text notes is a unique task on the other hand it is uniquely challenging task [34]. As mentioned earlier, our team hit second place in this competition. The first-place team utilized a rule-based method for all tasks, which narrowly defeated our team, although the scores were very close. The rule-based approach relies on human capabilities to identify label meanings and performs alignment using 'if-else' statements and other formulas in the program code. Due to the manual nature of labeling the text notes, it takes more time to comprehensively examine all the data compared to LLMs, which can also take many days. Moreover, these methods struggle to adapt when new types of data are introduced, which may decrease the overall performance. Therefore, in this work, the existing LLM method can match human capabilities in de-identifying EMR text notes and can alleviate the limitations of the traditional rule-based method.

5 Conclusion

This study aims to leverage the powerful capabilities of large language models for precise de-identification of patient information. This method not only ensures the protection of patient privacy but also expands the potential of electronic medical records for personalized medicine and innovative research. The proposed approach employs a sliding window strategy for data aggregation, explicitly identifying and managing minority labels to enhance the accuracy and efficiency of de-identification further. Additionally, we implemented a masking strategy to guide the model's attention towards generating the intended output results, thereby ensuring the reliability of privacy protection.

Furthermore, we introduced the NEFtune technique, designed to enhance the learning environment by comparing it with models of equal parameters, thereby improving training efficiency and optimizing the model's performance. This technique not only increases the model's ability to de-identify patient information but also enhances the overall system's reliability and accuracy.

However, our experimental results indicate that the performance of various large language models is not yet robust enough. This is primarily due to hardware resource limitations, which restrict our comparative analysis to models with outstanding performance and the latest architectures. This limitation points to an interesting direction for future research in this field, such as exploring more efficient hardware acceleration solutions or optimizing model architectures to improve performance and stability.

Acknowledgments. This work was supported by the 2024 International Workshop on De-identification of Electronic Medical Record Notes (IW-DMRN) (https://www.sredhconsortium. org/sredh-workshops/2024-iw-dmrn), organized by the SREDH Consortium (https://www.sredhc onsortium.org/sredh-consortium), National Kaohsiung University of Science and Technology, and Asian University. Additionally, we acknowledge the supervision and support provided by the Ministry of Education in Taiwan as part of the Ministry of Education Artificial Intelligence Competition and Annotation Data Collection Project.

Disclosure of Interests. All authors of this article declare that we have no competing interests related to the content of this article.

References

1. SREDH Consortium Homepage. https://www.sredhconsortium.org/. Accessed 16 Mar 2024
2. SREDH Consortium. 2024 International workshop on deidentification of electronic medical record notes (IW-DMRN). https://www.sredhconsortium.org/sredh-competitions/sredhai-cup-2023/2024-iw-dmrn
3. Codalab. Competition. https://codalab.lisn.upsaclay.fr/competitions/15425. Accessed 16 Mar 2024
4. Mir, T.H., et al.: Deidentification and temporal normalization of electronic health record notes using large language models: the SREDH/AI-Cup 2023 deidentification competition. In: 2024 International Workshop on Deidentification of Electronic Medical Record Notes, Springer Nature, Kaohsiung, Taiwan (2024)
5. Chen, A., Jonnagaddala, J., Nekkantti, C., Liaw, S.T.: Generation of surrogates for de-identification of electronic health records. In: MEDINFO 2019: Health and Wellbeing e-Networks for All, pp. 70–73 (2019)
6. Alla, N.L.V., Chen, A., Batongbacal, S., Nekkantti, C., Dai, H.J., Jonnagaddala, J.: Cohort selection for construction of a clinical natural language processing corpus. Comput. Methods Programs Biomed. Update **1**, 100024 (2021)
7. Nori, H., King, N., McKinney, S.M., Carignan, D., Horvitz, E.: Capabilities of GPT-4 on medical challenge problems. arXiv preprint arXiv:2303.13375 (2023)

8. Wang, S., Zhao, Z., Ouyang, X., Wang, Q., Shen, D.: ChatCAD: interactive computer-aided diagnosis on medical image using large language models. arXiv preprint arXiv:2302.07257 (2023). https://doi.org/10.48550/arXiv.2302.07257

9. Wang, Y., Ma, X., Chen, W.: Augmenting black-box LLMs with medical textbooks for clinical question answering. arXiv preprint arXiv:2309.02233 (2023)

10. Jin, M., et al.: Health-LLM: personalized retrieval-augmented disease prediction model. arXiv preprint arXiv:2402.00746 (2024)

11. Humerick, M.: Taking AI personally: how the EU must learn to balance the interests of personal data privacy & artificial intelligence. Santa Clara High Tech, LJ **34**, 393 (2017)

12. Price, W.N., Cohen, I.G.: Privacy in the age of medical big data. Nat. Med. **25**(1), 37–43 (2019)

13. Anom, B.Y.: Ethics of Big Data and artificial intelligence in medicine. Ethics Med. Public Health **15**, 100568 (2020)

14. Seh, A.H., et al.: Healthcare data breaches: insights and implications. Healthcare **8**, 133 (2020)

15. Kushida, C.A., Nichols, D.A., Jadrnicek, R., Miller, R., Walsh, J.K., Griffin, K.: Strategies for de-identification and anonymization of electronic health record data for use in multicenter research studies. Med. Care **50**, 82–101 (2012)

16. Yang, X., et al.: A study of deep learning methods for de-identification of clinical notes in cross-institute settings. BMC Med. Inform. Decis. Mak. **19**(5), 1–9 (2019)

17. Yan, R., Jiang, X., Dang, D.: Named entity recognition by using XLNet-BiLSTM-CRF. Neural Process. Lett. **53**(5), 3339–3356 (2021)

18. An, Y., Xia, X., Chen, X., Wu, F.X., Wang, J.: Chinese clinical named entity recognition via multi-head self-attention based BiLSTM-CRF. Artif. Intell. Med. **127**, 102282 (2022)

19. Naseem, U., Khushi, M., Reddy, V., Rajendran, S., Razzak, I., Kim, J.: BioALBERT: a simple and effective pre-trained language model for biomedical named entity recognition. In: 2021 International Joint Conference on Neural Networks (IJCNN), pp. 1–7. IEEE (2021)

20. Wang, S., et al.: GPT-NER: named entity recognition via large language models. arXiv preprint arXiv:2304.10428 (2023)

21. Dong, Q., et al.: A survey for in-context learning. arXiv preprint arXiv:2301.00234, (2022)

22. Hu, Y., et al.: Zero-shot clinical entity recognition using ChatGPT. arXiv preprint arXiv:2303.16416 (2023)

23. Shyr, C., Hu, Y., Harris, P.A., Xu, H.: Identifying and extracting rare disease phenotypes with large language models. arXiv preprint arXiv:2306.12656 (2023)

24. Goel, A., et al.: LLMs accelerate annotation for medical information extraction. Mach. Learn, Health (ML4H), 82–100 (2023)

25. Bian, J., Zheng, J., Zhang, Y., Zhu, S.: Inspire the large language model by external knowledge on biomedical named entity recognition. arXiv preprint arXiv:2309.12278 (2023)

26. Jonnagaddala, J., Chen, A., Batongbacal, S., Nekkantti, C.: The OpenDeID corpus for patient de-identification. Sci. Rep. **11**(1), 19973 (2021)

27. Bai, J., et al.: Qwen technical report. arXiv preprint arXiv:2309.16609 (2023)

28. Wang, L., et al.: Investigating the impact of prompt engineering on the performance of large language models for standardizing obstetric diagnosis text: comparative study. JMIR Formative Res. **8**(1) (2024)

29. Zhu, Y., et al.: REALM: RAG-driven enhancement of multimodal electronic health records analysis via large language models. arXiv preprint arXiv:2402.07016 (2024)

30. Zhao, Y., et al.: Pytorch FSDP: experiences on scaling fully sharded data parallel. arXiv preprint arXiv:2304.11277 (2023)

31. Jain, N., et al.: NEFTune: noisy embeddings improve instruction finetuning. arXiv preprint arXiv:2310.05914 (2023)

32. Wei, J., et al.: Chain-of-thought prompting elicits reasoning in large language models. Adv. Neural. Inf. Process. Syst. **35**, 24824–24837 (2022)

33. Liu, Y., Peng, X., Du, T., Yin, J., Liu, W., Zhang, X.: ERA-CoT: improving chain-of-thought through entity relationship analysis. arXiv preprint arXiv:2403.06932 (2024)
34. Liu, J., et al.: OpenDeID pipeline for unstructured electronic health record text notes based on rules and transformers: deidentification algorithm development and validation study. J. Med. Internet Res. **25**, e48145 (2023). https://doi.org/10.2196/48145

Applying Language Models for Recognizing and Normalizing Sensitive Information from Electronic Health Records Text Notes

Sheng-Xuan Huang[2]([📧]) [ID], Hung-An Cheng[2] [ID], and Zheng-Hao Li[1] [ID]

[1] Department of Electrical Engineering, College of Electrical Engineering and Computer Science, Taipei City, Taiwan

[2] National Kaohsiung University of Science and Technology, No. 415, Jiangong Road, Sanmin District, Kaohsiung City 807618, Taiwan R.O.C.
{F111154162,F112154142}@nkust.edu.tw

Abstract. Electronic medical records (EMRs) are often populated with private or confidential information related to patients. In the setting of the hospital environment, fragments of information could be collected across various electronic health record systems that could be used to deduce the true identities of patients mentioned in EMR text notes. Therefore, applying a de-identification procedure is a crucial means of safeguarding privacy, especially when handling EMR text notes. By preventing the identification of individuals and reducing the risks associated with personal information, it contributes to regulatory compliance, facilitates research and data sharing, and mitigates the risk of data misuse. In 2023, the Ministry of Education in Taiwan sponsored a large nationwide competition, AI-CUP 2023-privacy Protection and standardization of EMR Challenge to seek automatic de-identification and standardization solutions. As one of the participating teams, we first tried to apply zero-shot, one- shot, and few-shot configurations based on ChatGPT, but the preliminary experimental results showed that ChatGPT cannot follow the prompt to generate the prediction results of its recognized SHIs based on the official output format. Therefore, we decided to primarily utilize the Pythia-160m-deduped pre-trained model to develop our system. Through multiple experiments, the developed system achieved official micro- and macro-F-scores of 0.744356 and 0.5960788 respectively on the PPSEMR test set.

Keywords: Electronic medical records text notes · Sensitive health information · Pythia · ChatGPT · de-identification

1 Introduction

Electronic medical records (EMRs) [1] are essential tools used in healthcare institutions for storing and managing are commonly encountered in various fields, including signal processing, data analysis, and patient health information [2, 3]. This information may include sensitive data such as patient personal identification details, medical records, diagnostic results, prescription medications, and more. To safeguard patients'

J. Jonnagaddala et al. (Eds.): IW-DMRN 2024, CCIS 2148, pp. 72–86, 2025.
https://doi.org/10.1007/978-981-97-7966-6_6

privacy and ensure information security, healthcare institutions typically employ de-identification measures. This involves removing or obfuscating patient identification details (such as names, addresses, phone numbers, etc.) from medical records to ensure that only authorized personnel can access the actual identities associated with the data. This measure not only helps in compliance with regulations and standards but also aids in reducing the risk of data breaches that institutions may face.

De-identification is a privacy protection method designed to safeguard individual privacy when handling or sharing sensitive data. The process involves removing or modifying personally identifiable information in the data to reduce the risk. This includes eliminating names, addresses, identification numbers, and other personal identifiers to ensure that the data is no longer associated with specific individuals in a given context. De-identification helps ensure that data usage is lawful while protecting the privacy of participants.

Temporal normalization task refers to a process in which temporal data is standardized or normalized to a consistent format or scale. This task is based on machine learning, where time-related information needs to be aligned or adjusted for comparison or analysis purposes. The goal is to make temporal data comparable and consistent, often involving the alignment of timestamps, normalization of time intervals, or adjustment of temporal features. Temporal normalization is crucial for ensuring accurate and meaningful analysis of time-dependent data across different contexts or datasets.

To extract sensitive information from medical texts and standardize time information, we explore the possibility of directly applying ChatGPT as a means to recognized and normalize SHI information. Based on the preliminary analysis of the results of ChatGPT, our team decided to utilize the fine-tuning technique for language models (LLMs) based on the pre-trained Pythia model [4] implemented in the Hugging Face Transformers package [5] to address the artificial intelligence (AI) CUP 2023 privacy protection and standardization of EMR Competition (PPSEMR) [6–9]. The reasons for selecting the pre-trained Pythia [10] model are as follows:

Performance: The pre-trained Pythia model demonstrates strong performance in natural language processing tasks, with excellent language understanding and generation capabilities. Based on a preliminary analysis of the ChatGPT results, we believe the Pythia model provides a solid foundation for further fine-tuning to address the challenges of AI CUP 2023 PPSEMR.

Compatibility: The Pythia model has already been implemented in the Hugging Face Transformers package, making it easy to integrate with our workflow and existing technological stack. This facilitates convenient and efficient fine-tuning using the model.

Community Support: The Hugging Face Transformers package is a widely used open-source toolkit with a large community of support and contributions. This means that we can leverage resources, documentation, and tutorials provided by the community to better understand and apply the Pythia model and to receive support when needed.

Overall, based on its performance, compatibility, and community support, we have decided to utilize the pre-trained Pythia model for fine-tuning to address the privacy protection and EMR standardization challenges of AI CUP 2023 PPSEMR. The proposed method jointly addresses the tasks of SHI recognition and normalization and gets rid of the offset errors observed in the output of ChatGPT. The developed system achieved official micro- and macro-F-scores of 0.744356 and 0.5960788 respectively on the PPSEMR test set.

2   Methods

2.1   Preliminary Study of Using ChatGPT

We describe our preliminary study of applying zero-shot, one-shot, and few-shot methods to prompt ChatGPT to recognize sensitive health information (SHI) and generate its output following the official output format illustrated in Fig. 1. The output consists of six fields: the file ID, the SHI type, the starting offset, the ending offset, the text of the SHI and its corresponding normalized value if the recognized SHI is date-related.

```
9 DATE 8 18 09/08/2957 2957-08-09
9 TIME 38 57 14/02/3014 at 11:42 3014-02-14T11:42
9 DATE 208 214 Friday 3014-02-18
9 DEPARTMENT 69 85 3 ARRIETTA CLOSE
9 HOSPITAL 87 89 POW
9 DOCTOR 93 108 AADLAND ABRAHAM
```

Fig. 1. The official output defined in the AICUP-23-PPSEMR challenge.

In this context, zero-shot refers to providing new input to a model without prior training on specific domains, and without giving corresponding examples or prompts. The aim is to observe the model's performance on unseen data without relying on previous domain-specific training. In natural language processing, this may involve presenting a sentence or question, and the model is expected to generate an answer directly without prior examples. The zero-shot approach is often used to assess a model's generalization ability and adaptability to new contexts.

One-shot refers to the training of a model with only one example or prompt, allowing it to make predictions on new inputs. This approach aims to assess the model's learning and generalization capabilities with minimal data, emphasizing the ability to learn from and predict with a very small number of examples compared to traditional multi-example training.

Few-shot refers to training a model with a very small number of data examples or prompts, enabling it to make predictions on new inputs. This approach emphasizes learning and generalization capabilities with significantly fewer examples compared to traditional multi-example training. In contrast to one-shot, few-shot allows for the use of a slightly larger number of examples but still much fewer than conventional training methods.

We conducted experiments to examine whether we can use prompting to guide ChatGPT to recognize SHIs mentioned in the training set released by PPSEMR challenge and output in according to the desired format. The results of using the zero- shot configuration are depicted in Fig. 2, indicating that although ChatGPT can recognize some correct SHI, it fails to produce the correct format and generates irrelevant output.

```
file31  IDNUM        8      18     481336392R
file31  MEDICALRECORD 27     34     4833363
file31  HOSPITAL     46     67        ROYAL HOBART HOSPITAL
file31  IDNUM        142    152    481336392R
file31  MEDICALRECORD 153    164    4833363.XHR
file31  PATIENT      166    180    HERBER, VERNON
file31  IDNUM        191    199    48133639
file31  IDNUM        200    208    48133639
file31  STREET       209    217    Rippling
file31  CITY         218    223    YOUNG
file31  STATE        225    227    NT
file31  ZIP          229    233    2008
file31  DATE         266    275    1/10/1996          1996-10-01
file31  TIME         295    314    17/10/2012 at 13:27    2012-10-17T13:27
file31  DEPARTMENT   326    343    7.ACUTE CARE WARD
file31  HOSPITAL     344    359       INJUNE HOSPITAL
file31  DOCTOR       363    382    ZAYA KINA REINHOLDT
file31  DOCTOR       485    499    MARVIN DERENZI
file31  DATE         531    541    19/12/2062         2062-12-19
file31  IDNUM        559    569    481336392R
file31  DATE         2001   2011   28/12/2062         2062-12-28
file31  DOCTOR       2037   2050   ROSCOE GEMMEN
file31  IDNUM        2219   2229   481336392R
file31  DATE         3384   3392   25/12/62           2062-12-25
file31  DOCTOR       3416   3430   CORTEZ SCULLIN
file31  DOCTOR       3455   3467   QUINN MEILER
```

```
plaintext                                          Copy code

楷名: file31
分類: Pathology Report
關鍵詞:
1. SPR no: 6SB044723A
2. MRN no: 6500447
3. Site_name: MOREE DISTRICT HOSPITAL
4. Facility_id: 016
5. Specimen_type: Fresh Tissue
6. Lab No: 6SB04472
7. Specimen: Marrow, Tissue
8. D.O.B: 29/10/1948
9. Sex: M
10. Collected: 17/12/2014 at 13:27
11. Location: Envoi-PENOLA WAR MEMORIAL HOSPITAL
12. Clinical History/Indication: Review diagnosis of multiple myeloma.
13. Name: JARRED KOSINSKI
14. MRN: 1109704
15. Date performed: 17/12/2014
16. Episode number: 40N225472N
17. Procedure: BMAT
18. Site: R PSIC
19. Peripheral Blood:
    - Hb: 123g/L
    - Blasts: 0.0×10e9/L
    - WBC: 3.47×10e9/L
    - Monocytes: 0.3×10e9/L
    - Platelets: 150×10e9/L
    - Eosinophils: 0.08×10e9/L
    - Neutrophils: 1.7×10e9/L
```

Fig. 2. On the left are the correct answers for "file31," while on the right are the results from ChatGPT using the zero-shot approach.

We further try to use one-shot setting resulting in the output of desired format but there are some unwanted outputs such as "ORGAN" and "TUMORTYPE" shown in Fig. 3. Finally, we utilized the few-shot approach and made prompt adjustments based on our error analysis shown in Fig. 4. The results showed that, aside from the errors of the starting/ending offsets, ChatGPT could recognize SHIs along with the normalized values. Our preliminary estimation for the performance of few-shot ChatGPT configuration on a validation set sampled from the training set is [0,0] for subtasks 1 and 2, respectively.

```
9   MEDICALRECORD 1     11    916204.JEL
9   PATIENT       12    26    VADENAIS, TROY
9   IDNUM         27    35    91R20431
9   DATE          51    60    10/3/2015        2015-03-10
9   DOCTOR        1327  1329  YA
9   DOCTOR        1330  1332  YA
9   DATE          1336  1344  20.11.15         2015-11-20
9   DOCTOR        1375  1385  Y Vadenais
9   DOCTOR        2235  2245  Y Vadenais
9   DATE          2249  2259  01.12.2015       2015-12-01
9   DOCTOR        2308  2318  Y Vadenais
```

```
plaintext                                          Copy code

楷名2:
file9  IDNUM        1    10   916204.JEL
file9  PATIENT      11   25   VADENAIS, TROY
file9  MEDICALRECORD 26   34   91R20431
file9  DATE         77   92   10/3/2015    2015-10-03
file9  DOCTOR       97   115  Dr Y Vadenais
file9  ORGAN        165  174  Prostate
file9  TUMORTYPE    175  196  Acinar adenocarcinoma
file9  GLEASONSCORE 197  213  3 + 4 = 7
file9  STAGE        214  229  pT2b pN0 MX
file9  MARGINS      230  242  Margins clear.
file9  INVASION     243  266  Perineural invasion: Present
file9  LYMPHNODES   267  303  A total of 8 negative lymph nodes (3 right
file9  ADDITIONAL   354  503  The relevant slides have been reviewed with
```

Fig. 3. On the left are the correct answers for "9", while on the right are the results from ChatGPT using the one-shot approach.

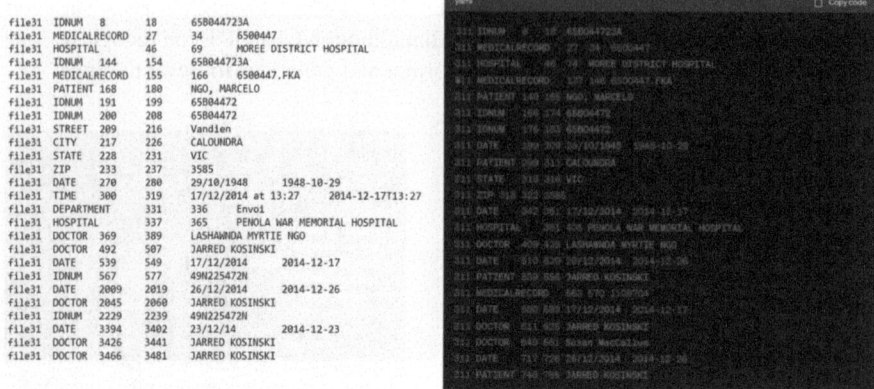

```
file31 IDNUM      8     18    658044723A
file31 MEDICALRECORD 27   34    6500447
file31 HOSPITAL   46    69    MOREE DISTRICT HOSPITAL
file31 IDNUM    144    154   658044723A
file31 MEDICALRECORD 155  166   6500447.FKA
file31 PATIENT  168    180   NGO, MARCELO
file31 IDNUM    191    199   65804472
file31 IDNUM    200    208   65804472
file31 STREET   209    216   Vandien
file31 CITY     217    226   CALOUNDRA
file31 STATE    228    231   VIC
file31 ZIP      233    237   3585
file31 DATE     270    280   29/10/1948    1948-10-29
file31 TIME     300    319   17/12/2014 at 13:27   2014-12-17T13:27
file31 DEPARTMENT 331  336   Envoi
file31 HOSPITAL  337   365   PENOLA WAR MEMORIAL HOSPITAL
file31 DOCTOR   369   389   LASHAWNDA MYRTIE NGO
file31 DOCTOR   492   507   JARRED KOSINSKI
file31 DATE     539   549   17/12/2014    2014-12-17
file31 IDNUM    567   577   49N225472N
file31 DATE     2009  2019  26/12/2014    2014-12-26
file31 IDNUM    2229  2239  49N225472N
file31 DATE     3394  3402  23/12/14      2014-12-23
file31 DOCTOR   3426  3441  JARRED KOSINSKI
file31 DOCTOR   3466  3481  JARRED KOSINSKI
```

Fig. 4. On the left are the correct answers for "file31", while on the right are the results from ChatGPT using the few-shot approach.

2.2   Language Model-Based Fine Tuning

We have concluded that fine-tuning the language model is the most effective strategy for winning the PPSEMR competition. This conclusion was reached after considering the expense of modifying ChatGPT and the fact that the offset information needed to be more accurate. Mentioned below is an explanation of the strategies that we have used inside the subsections that are to come.

Dataset Pre-processing. For a given pathology report, we utilized sentence segmentation tool to segment sentences. The splitted sentence content is extracted to form the format shown in Fig. 5. The last column is the content part which is linked with the gold annotations. The redundant content is filtered to reduce the size of the training set.

```
Before                                          After
Episode No:  88Y206206L                  1001   0     Episode No:  88Y206206L
8892062.BPL                              1001   24    8892062.BPL
                                         1001   37    "Vatterott, Jerrie CLARENCE"
Vatterott, Jerrie CLARENCE               1001   65    "Lab No:  88Y20620,88Y20620"
Lab No:  88Y20620,88Y20620               1001   92    Exeter
Exeter                                   1001   99    DECEPTION BAY  Northern Territory  6845
DECEPTION BAY  Northern Territory  6845  1001   139   "Specimen: Fluid,Tissue"
Specimen: Fluid,Tissue                   1001   162   D.O.B:  15/11/2004
D.O.B:  15/11/2004                       1001   181   Sex:  F
Sex:  F                                  1001   189   Collected: 20/5/2064 at :
Collected: 20/5/2064 at :                1001   215   Location:  PARKES 8 - GUNNEDAH DISTRICT HOSPITAL
Location:  PARKES 8 - GUNNEDAH DISTRICT HOSPITAL 1001 264   DR Edison Clay GOLDHIRSH
DR Edison Clay GOLDHIRSH                  1001  289   "Distribution:   FILE-COPY,   NSW-CANCER-REGISTRY"
Distribution:   FILE-COPY,   NSW-CANCER-REGISTRY
```

Fig. 5. The dataset after preprocessing. The first column is the file ID, the second column is the sentence ID, and the last column is the corresponding content.

Applying Language Model for the AICUP 2023 PPSEMR Subtasks. We used the pythia-70m, 160m, and 410m models as the fundamental language models (LM) to address the challenge. The release of the Pythia model suite is to facilitate researchers to investigate the behavior, functionality, and limitations of LLMs. The Pythia suite is designed to provide a controlled setup for scientific experiments offering 154 checkpoints for each model size including the initial step, 10 checkpoints at logarithmic intervals {1, 2, 4... 512}, and 143 evenly spaced checkpoints from step 1000 to step 143000. These checkpoints are hosted as branches on Hugging Face. In our experiment [11], the Pythia model trained at step 3000 was used. We employed AdamW along with a learning rate warm-up for adjusting the learning rate with an initial value of $5e^{-5}$. The batch size is set to 5 due to hardware limitations; exceeding 5 would exceed the memory constraint. The batch size is set to 5 due to hardware limitations, which would exceed the memory limit. The checkpoint at 10 refers to the 10th epoch. The choice of using checkpoint 10 is because, based on the loss values during the training process as shown in Fig. 6, the loss curve has already become relatively flat around the 10th epoch. Additionally, according to the scoring algorithm calculating the test scores for each checkpoint, checkpoint 10 performed the best. Further increasing the number of epochs not only does not significantly improve the score but also increases the training time required.

Fig. 6. After individual training for different numbers of times, the loss value for each training session is calculated. It can be observed that by the 10th training session, the loss curve has already flattened out.

Similar to the results illustrated in the ChatGPT experiment, the much smaller Pythia 160M model did not comprehend the structure of the output format well either. To address this, we therefore define a prompt format as shown in Fig. 7. Along with post-processing procedures to (1) extract the recognized SHIs and corresponding normalized values (the {ANNOTATIONS} part of the template), (2) remove duplicated outputs generated by LLM, (3) non-relevant SHI types which are not considered in the competition, and (4) non-exist SHIs which are hallucinated by LLM; the errors can be identified if we cannot find the recognized SHI's offset from the input context.

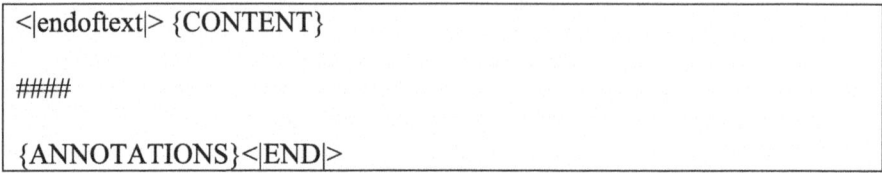

Fig. 7. The prompt format defined for the fine-tuning procedure.

The training configuration of our method is summarized as follows.

The hardware configuration is a computer equipped with a CPU (Intel i5 136000k) and GPU (NVIDIA GeForce 3060ti). The operating system used is Windows 11 installed with the following Python packages: Transformers, PyTorch, Datasets, Random, PEFT, TQDM, and Matplotlib. Python is employed as the primary programming language, with the PyTorch framework used for constructing and training the LM models. The models are run on GPU for accelerated computations. We used the tools and LM released in the Transformers package for fine-tuning the pre-trained models. In particular, the Datasets package is used for convenient handling and loading of the PPSEMR datasets. Python's built-in Random library is used for randomly initializing the parameters, the PEFT library is used for performance evaluation and training along with TQDM for displaying progress bars, and Matplotlib for data visualization.

3   Results

In our experiment, we first utilized Pythia-70 m in the fine-tuning procedure. For sub-task 1, we obtained a low macro-F-score of 0.291, and for subtask 2, we achieved an even lower macro-F-score of 0.192. Subsequently, we decided to use Pythia-160 m, which improved the macro-F-score of our system to 0.328 and 0.412 for subtasks 1 2, respectively. After analyzing the models' output and hyperparameter tuning, we tested with both Pythia-160 m and Pythia-410 m models. The results showed that after fine-tuning, the performance of these two models was relatively similar, but the computation time for the 410m model significantly exceeded the competition's specified time limit. Therefore, we decided to abandon the use of the 410m model.

Based on our experiments on the validation set, we found that the highest macro-F-scores could be achieved at the 12th training iteration, as shown in Fig. 8. Therefore, we set the number of training iterations to 12 in our implementation. During this period, we also experimented with incorporating LoRA [12–14]. Although the inclusion of LoRA

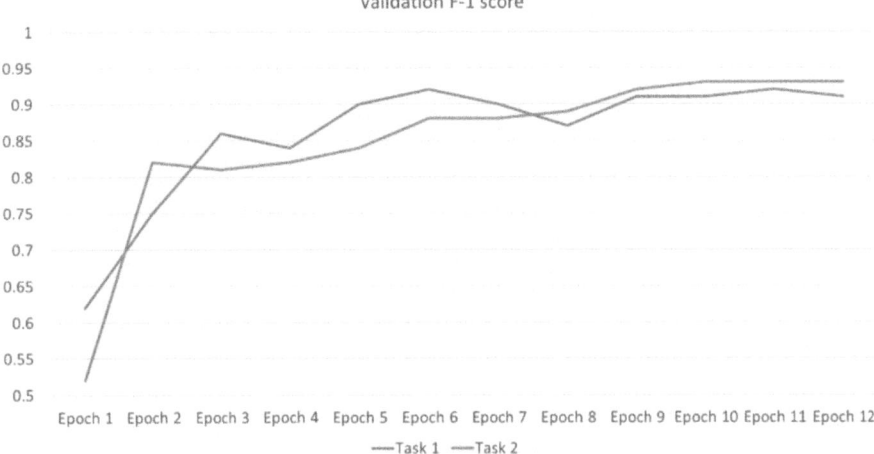

Fig. 8. Performance comparison of macro-F-scores for each iteration of subtasks 1 and 2 using validation dataset.

can significantly reduce the training time, the scores were lower. Ultimately, we chose Pythia-160 m as our main model to optimize its performance.

Through the experiments conducted during the completion, we noticed that the model underperformed in certain SHI types which contain less annotations in the released training set. To address this, we implemented a subroutine to save the weights of the model after each training iteration. According to the results of the few-shot approach introduced in the previous section for ChatGPT's ability to generate synthetic data, we used the prompt shown in Fig. 9 to ask ChatGPT to generate additional training instances to augment the lower-scoring SHI types observed during the comparison at each training iteration. After incorporating the augmented training set, there was a significant improvement for the Pythia-160m model as illustrated in Fig. 10 and Table 1 on the pre-validation set; the Pythia-160m model can achieve a high score of 0.918 for subtask 1 and 0.936 for subtask 2.

Table 1. Official validation set performance.

Pythia-70m	Pythia-160m	Augmented-Pythia-60m
subtask 1: 0.291	Subtask 1: 0.328	Subtask 1: 0.918
subtask 2: 0.192	Subtask 2: 0.412	Subtask 2: 0.936

The Official validation utilized for fine-tuning our model is depicted in Fig. 11. Incorporating the augmented training set leads to a significant enhancement in performance, as evidenced by the results in Table 2 and Table 3.

You

幫我用類似的語法 創造30條不一樣但格式一樣詞

89   4301   "I couldn't find any New Zealand facilities offering MDR3 gene analysis."
COUNTRY: New Zealand

902   116   "[High-grade squamous cell carcinoma with clear margins assessed by Dr.
Emma LINCOLN, AUCKLAND NZ]."   DOCTOR: Emma LINCOLN\nCITY:
AUCKLAND\nCOUNTRY: New Zealand

234   4523   "The diagnosis aligns with the initial assessment (identified as high-grade
squamous cell carcinoma by Dr. Adam WATSON in BRISBANE, AUSTRALIA)."
DOCTOR: Adam WATSON\nCITY: BRISBANE\nCOUNTRY: Australia

ChatGPT

Sure, here are 30 variations with similar syntax:

. 567 8901 "No facilities in Japan offer analysis of the MDR3 gene, as I discovered."
  COUNTRY: Japan

. 342 672 "[Clear margins evaluated by Dr. Sakura YAMAMOTO in TOKYO, JAPAN,
  indicate high-grade squamous cell carcinoma]."

  DOCTOR: Sakura YAMAMOTO   ↓

  CITY, TOKYO

Fig. 9. Prompts used for generating additional training instances based on ChatGPT.

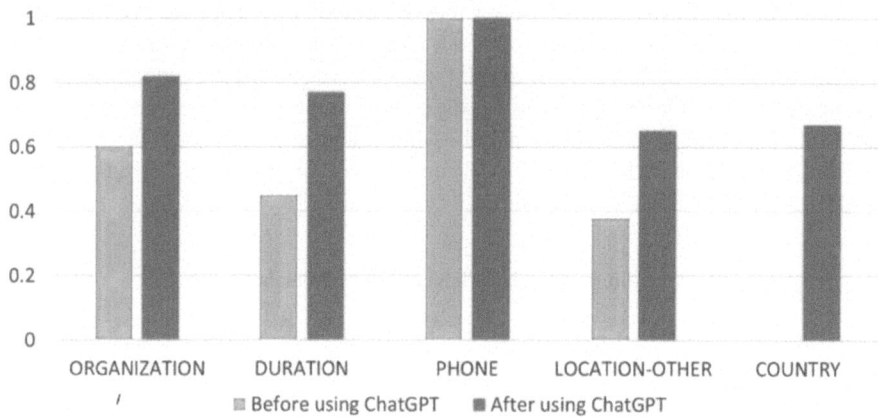

Fig. 10. Official validation set performance comparison for the models trained on the original and augmented datasets.

Table 2. Official validation TASK1

Coding Type	Precision	Recall	F-measure	Support
MEDICALRECORD	0.9633508	0.9804618	0.971831	563
PATIENT	0.7364975	0.8256881	0.7785468	545
IDNUM	0.9707149	0.9575191	0.9640719	1177
DATE	0.9693558	0.9615384	0.9654313	1612
DOCTOR	0.8488972	0.8256504	0.8371124	2191
STREET	0.7854251	0.6043614	0.6830986	321
CITY	0.978395	0.9406528	0.9591529	337
STATE	0.9804561	0.9709678	0.9756888	310
ZIP	0.8401254	0.8246154	0.8322981	325
DEPARTMENT	0.9594203	0.9043716	0.931083	366
HOSPITAL	0.9589744	0.9460371	0.9524618	593
DURATION	0.6	0.5	0.5454546	6
TIME	0.9005376	0.8014354	0.8481013	418
AGE	0.9622642	0.8947368	0.9272727	57
ORGANIZATION	0.6363636	0.5833333	0.6086956	24
LOCATION-OTHER	1	0.25	0.4	4
PHONE	1	1	1	2
COUNTRY	0	0	0	2
Micro-avg. F	0.9086289	0.8885124	0.898458	8853
Micro-avg. F	0.8383766	0.7650761	0.8000509	8853

Table 3. Official validation TASK2

Temporal Type	Precision	Recall	F-measure	Support
DATE	0.8193548	0.7878412	0.8032891	1612
DURATION TIME	1	0.5	0.6666667	6
Micro-avg	0.9343284	0.7488039	0.8313414	418
Micro-avg	0.8400424	0.7789784	0.8083588	2036
	0.9178944	0.6788816	0.7804998	2036

In the final competition dataset, we present the scores for each label in subtask 1 and subtask 2, as shown in Figures 12 and 13. More detailed data can be found in Tables 4 and 5. By comparing Tables 2 and 3, we observe significant decreases in scores for "TIME," "ORGANIZATION," and "PHONE" in subtask 1, and particularly dismal scores for "TIME" in subtask 2. We will further rectify these errors and investigate the reasons behind such substantial score drops.

In the future, we will consider establishing a model for preprocessing the dataset because there are still many errors in the preprocessing data provided by the competition organizer. Before this, we manually corrected and modified it, which consumed a lot of time, and there is still a possibility of errors in manual correction. Regarding training, we want to abandon pretrained models because we found that regardless of the pretrained model used, it may still generate words or gibberish that are not within the selected

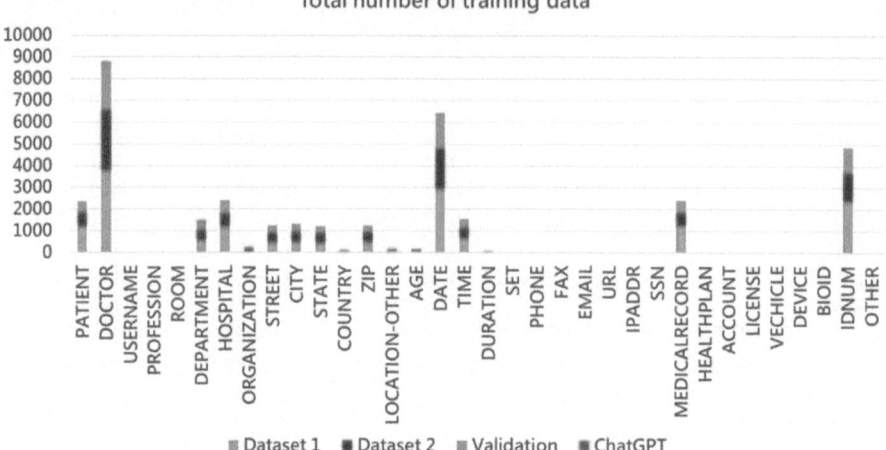

Fig. 11. Number of labels in the training set and validation set.

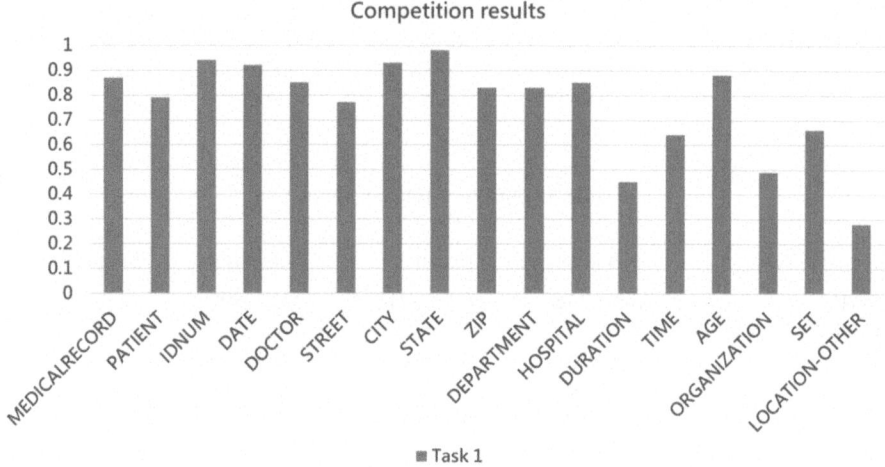

Fig. 12. Scores for various labels in Task 1 of the competition set.

labels. Although pretrained models are powerful, fine-tuning is still required to meet our needs. By establishing a language training model specifically for EMR text notes, we can avoid a large number of tedious fine-tuning steps, thereby allowing untrained healthcare personnel to use it easily.

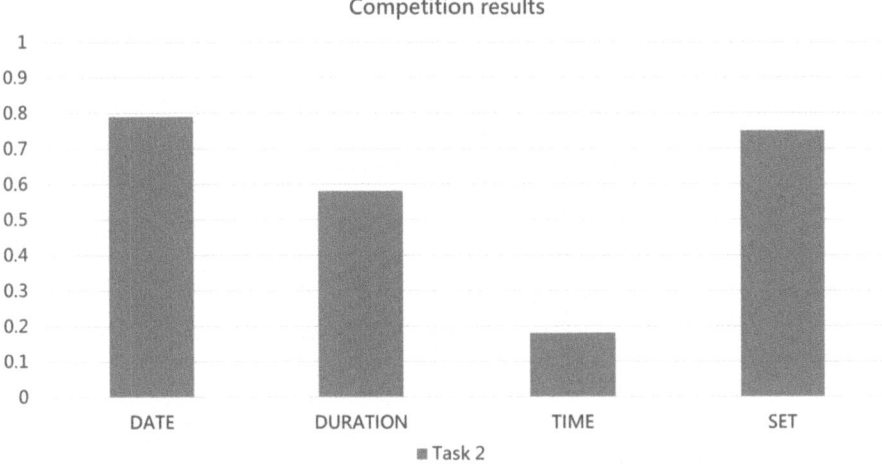

Fig. 13. Scores for various labels in Task 2 of the competition set.

Table 4. The final performance on subtask 1.

Coding Type	Precision	Recall	F-measure	Support
MEDICALRECORD	0.7817796	0.9879518	0.8728563	747
PATIENT	0.7439024	0.8519553	0.7942708	716
IDNUM	0.9434844	0.9528302	0.9481342	2120
DATE	0.90887024	0.9511997	0.9294655	2459
DOCTOR	0.8808538	0.8310791	0.8552428	3327
STREET	0.9126984	0.6686047	0.771812	344
CITY	0.9530387	0.924933	0.9387755	373
STATE	0.987842	0.9789157	0.9833586	332
ZIP	0.8362069	0.8302425	0.8302425	353
DEPARTMENT	0.8746736	0.7995227	0.8354115	419
HOSPITAL	0.8664383	0.8447412	0.8554522	1198
DURATION	0.5	0.4166667	0.4545455	12
TIME	0.84375	0.5170213	0.641161	470
AGE	0.9761904	0.8039216	0.8817204	51
ORGANIZATION SET	0.4505495	0.5540541	0.4969697	74
LOCATION-OTHER PHONE	0.75	0.6	0.6666667	5
Micro-avg. F	1	0.1666667	0.2857143	6
Micro-avg. F	0	0	0	1
	0.8796526	0.8721458	0.8758831	13007
	0.7894505	0.7041348	0.744356	13007

Table 5. The final performance on subtask 2.

Temporal Type	Precision	Recall	F-measure	Support
DATE	0.8199316	0.7795852	0.7992495	2459
DURATION TIME	1	0.4166667	0.5882353	12
SET	0.2757202	0.1425532	0.1879383	470
Micro-avg. F	1	0.6	0.75	5
Micro-avg. F	0.7694091	0.6761711	0.7197832	2964
	0.7739129	0.4847013	0.5960788	2964

4   Discussion

In this study, we found that our final scores were not as ideal compared to the training scores, especially in the aspect of time normalization for Task 2. To address this issue, we plan to first modify the prompts for Task 2. For both Task 1 and Task 2, we will adjust the training parameters separately to achieve a target score of 90, while maintaining minimal hardware requirements and the fastest training time possible. We also intend to utilize more advanced models for training if possible. The main challenge we encountered in this study was the inadequate hardware equipment. We believe that with better equipment and optimized code, we can predict results more quickly and accurately. Before that, we expect to achieve the desired results using lower-spec equipment and faster training speeds. The findings make it very clear that our research has potential for development. Although our final ratings were significantly lower, the data we obtained provided valuable insights and recommendations for prospective studies. Regarding Task 2, the problem of time normalization is one aspect that calls for more consideration. Even though we intend to tweak the prompts and adjust the training settings to enhance the scores, we are also aware of the fact that more sophisticated models and algorithms are required to solve this difficulty. In light of this, we suggest that future research investigate a variety of concepts and approaches to obtain more favorable outcomes in this domain. Another essential aspect to consider is the influence of hardware equipment on the investigation results. Because we had a limited number of resources, we were forced to make concessions in terms of both our speed and our precision.

Nevertheless, we can produce more accurate findings in less time if we have better equipment and optimized code. In addition, we suggest that future research with comparable aims prioritize the update of hardware to guarantee the highest possible level of performance. This research's findings emphasize the significance of consistently improving the models and methods used in machine learning. As technology improves, researchers must modify their methods and investigate new methodologies to produce better findings. We hope our results will invigorate more studies and contribute to the continuous effort.

5 Conclusion

To develop our system for the PPSEMR competition, our team has Pythia-160m at our disposal. After completing several iterations, we found that the Pythia-160M model, which had been fine-tuned in the 12th iteration, was the best one, in addition to an enhanced training dataset. On the other hand, we realized that efficient dataset pretreatment and postprocessing are necessary to assemble high-quality training datasets for the Language Models (LM) fine-tuning procedure. In order to reach the macro-F-scores of 0.744356 and 0.5960788 for subtasks 1 and 2, respectively, which are considered adequate, we recommended tweaks and optimizations to increase the performance of our model. We could show via the experiment results that the suggested tweaks and optimizations improved our model's capacity to comprehend and appropriately respond to the questions asked throughout the competition. Because of the commitment and effort that our team put in, we were able to achieve the 26th position in the tournament. We are very proud of our successes, and we look forward to putting our knowledge to use and further developing our approach.

Acknowledgments. The author would like to express gratitude to the teachers and institutions who contributed to this study. Special thanks are extended to Professor Hong-Jie Dai and his research team for their guidance in the initial programming phase. Additionally, gratitude is extended to the Ministry of Education for organizing the competition, and to the Department of Electrical Engineering at the National Kaohsiung University of Science and Technology (Intelligent Systems Laboratory), the Department of Biomedical Informatics and Medical Engineering at Asia University, and the SREDH (https://www.sredhconsortium.org/) Association of Australia for providing the competition topics.

Disclosure of Interests.   The authors have no competing interests.

References

1. Liu, J., et al.: OpenDeID pipeline for unstructured electronic health record text notes based on rules and transformers: deidentification algorithm development and validation study. J. Med. Internet Res. [Internet] **25**(1), e48145 (2023). https://www.jmir.org/2023/1/e48145
2. Bates, D.W., Mark, E., Edward, G., Zapp, J., Mullins, H.C.: A proposal for electronic medical records in U.S. primary care. J. Am. Med. Inform. Assoc. [Internet]. **10**(1), 1–10 (2003). https://doi.org/10.1197/jamia.M1097
3. Shortliffe, E.H.: The evolution of electronic medical records. Acad. Med. **74**(4), 414–419 (1999)
4. Uzuner, Ö., Luo, Y., Szolovits, P.: Evaluating the state-of-the-art in automatic de-identification. J. Am. Med. Inform. Assoc. [Internet]. **14**(5), 550–563 (2007). https://doi.org/10.1197/jamia.M2444
5. Hutter, F., Loshchilov, I.: Decoupled weight decay regularization. In: International Conference on Learning Representations, December 2018
6. SREDH [Internet]. www.sredhconsortium.org, https://www.sredhconsortium.org/. Cited 19 Apr 2024
7. SREDH - 2024 IW-DMRN [Internet]. www.sredhconsortium.org, https://www.sredhconsortium.org/sredh-competitions/sredhai-cup-2023/2024-iw-dmrn. Cited 19 Apr 2024

8. CodaLab - Competition [Internet]. codalab.lisn.upsaclay.fr. https://codalab.lisn.upsaclay.fr/competitions/15425. Cited 19 Apr 2024

9. Jonnagaddala, J., Chen, A., Batongbacal, S., Nekkantti, C.: The OpenDeID corpus for patient de-identification. Scientific Reports [Internet] **11**(1), 19973 (2021). https://www.nature.com/articles/s41598-021-99554-9

10. Biderman, S., et al.: Pythia: A Suite for Analyzing Large Language Models Across Training and Scaling [Internet], pp. 2397–2430. Proceedings.mlr. Press. PMLR (2023). https://proceedings.mlr.press/v202/biderman23a.html

11. Lester, B., Al-Rfou, R., Constant, N.: The power of scale for parameter-efficient prompt tuning. arXiv: 210408691 [cs] [Internet], 2 September 2021. https://arxiv.org/abs/2104.08691

12. Hu, E.J., et al.: LoRA: low-rank adaptation of large language models. In: International Conference on Learning Representations, October 2021

13. Chen, A., Jonnagaddala, J., Nekkantti, C., Liaw, S.T.: Generation of Surrogates for De-Identification of Electronic Health Records [Internet], p. 70-3. ebooks.iospress.nl. IOS Press (2019). https://ebooks.iospress.nl/publication/51950. Cited 19 Apr 2024

14. Mir, T.H., et al.: Deidentification and temporal normalization of electronic health record notes using large language models: the SREDH/AI-Cup 2023 deidentification competition. In: 2024 International Workshop on Deidentification of Electronic Medical Record Notes (2024). Kaohsiung, Taiwan: Springer Nature

Enhancing SHI Extraction and Time Normalization in Healthcare Records Using LLMs and Dual-Model Voting

PeiWen Huang[(✉)] [iD] and Tzu-En Liu[iD]

National Taiwan University, No. 1, Sec. 4, Roosevelt Road, Taipei 10617, Taiwan
d12922004@ntu.edu.tw, hane0131@gmail.com

Abstract. Our study evaluates the effectiveness of discriminative and generative models in protecting privacy and standardizing medical data, focusing on Sensitive Health Information (SHI) extraction and Time Information Normalization (TIN) in medical records. Utilizing advanced pre-trained models, we introduce a dual-model fusion with a voting mechanism, diverse data forms, and data augmentation techniques. Our experiments demonstrate the superiority of generative models over traditional discriminative models in the SHI extraction task. This paper underscores the significance of strategic model selection, adept data processing, and the novel voting mechanism in enhancing the performance of SHI extraction and TIN tasks. Our methods achieve remarkable performance both on SHI extraction and TIN tasks. These results pave the way for advanced AI applications in healthcare, offering valuable insights for ongoing research in medical data management and privacy safeguarding.

Keywords: Generative Models · Name Entity Recognition · Voting Mechanism · Sensitive Health Information Extraction · Time Information Normalization

1 Introduction

The AI-Cup 2023 [1] represents a critical initiative to advance privacy protection and medical data standardization in the era of digital health. This competition addresses the urgent need for automatic de-identification and standardization solutions in the face of privacy concerns raised using Large Language Models (LLMs) in clinical medicine and the pervasive use of Electronic Health Records text notes (EHRs). By focusing on the precise extraction of SHI and the normalization of time-related data within medical records, the AI-Cup 2023 aims to mitigate the risk of confidential information leakage and promote the safe use of EHR text notes for secondary research. Participants were challenged to evaluate their AI models on a large Australian multicenter corpus, with a primary evaluation on sensitive health information (SHI) entity recognition and time information normalization.

LLMs have emerged as a transformative force in healthcare data augmentation, exemplified by their sophisticated capabilities in improving patient-trial matching [3].

© The Author(s), under exclusive license to Springer Nature Singapore Pte Ltd. 2025
J. Jonnagaddala et al. (Eds.): IW-DMRN 2024, CCIS 2148, pp. 87–106, 2025.
https://doi.org/10.1007/978-981-97-7966-6_7

This reflects the broader applicability of LLMs in healthcare, where they can address complex tasks. Similarly, the domain of Named Entity Recognition (NER) has undergone significant evolution. Recent methodological reviews have charted the transition from rule-based to deep learning approaches, which has improved the precision of extracting critical information from EHR text notes [4]. These advancements have informed the methodologies we applied in our study for the SHI extraction and time normalization tasks. For the SHI extraction task, we utilize both traditional discriminative and the increasingly popular generative approaches. For the time information normalization task, we employ a generative model enhanced with regular expression pattern matching.

Main contributions to our study:

- We apply advanced pre-trained models for SHI extraction, encompassing both discriminative and generative approaches.
- Our study introduces a novel Dual-Model Fusion with a Voting Mechanism. This mechanism can preserve the strengths of auto-regressive and auto-encoder models to achieve optimal performance.
- We conduct a variety of methods to process data, including the utilization of concatenation, cropping, and data augmentation, particularly for smaller datasets.
- We enhance our models' learning efficacy with advanced training techniques, such as parameter freezing and Low-Rank Adaptation (LoRA [5]).

Overall, our study makes a significant stride forward in employing LLMs for the secure and efficient management of sensitive medical data. We provide a robust solution for data privacy and standardization challenges in healthcare by introducing innovative AI techniques.

2   Method

2.1   Task Definitions

This competition [1, 25] is centered around two interrelated tasks aimed at advancing privacy protection and data standardization in healthcare records, as detailed in the CodaLab competition's official website [6]. For our project resources, refer to our GitHub page [7].

Task 1: Protected Structured Health Information Extraction. Task 1 is focused on the extraction and classification of SHI from EHRs, a task that faces significant challenges such as the need for large and high-quality annotated datasets, and the complexity of data cleaning and annotation. Drawing inspiration from pioneering studies like Med-BERT [8], which showcases the application of NLP techniques on structured HER text notes to enhance disease prediction accuracy through contextualized embeddings, our objective is to precisely identify and categorize SHI entities such as names, locations, dates, and contact information. This initiative involves leveraging both discriminative and generative models to determine their efficacy in this domain, with a primary aim of meticulously extracting these sensitive pieces of information in alignment with competition standards, ensuring both thoroughness and adherence to predefined categories.

Task 2: Time Information Normalization. Task 2 is a relatively straightforward endeavor, following the extraction of SHI from EHR text notes. It focuses on the normalization of time information, specifically targeting four key categories identified within the extracted data: DATE, TIME, DURATION, and SET. By aligning this information with the ISO 8601 standard, Task 2 aims to ensure that dates and times are represented in a clear and universally understandable format that benefits both human interpretation and computer processing. This standard emphasizes a year-first format for dates and includes detailed time information such as hours, minutes, seconds, and milliseconds, along with time zone specifics [9]. Given the relative simplicity of this task, our approach primarily utilizes a generative model, specifically the flan-t5-base, which is further enhanced with regular expressions. This methodological choice facilitates the transformation of each identified time-related entity within the records into a consistent and standardized format, thereby achieving uniformity and clarity in the representation of time data across different systems.

2.2 Discriminative Approach

In Task 1, we approach the challenge as a Sequence Labeling problem. Given a string consisting of a sequence of words, our model is tasked with predicting a label for each word. These labels are predefined SHI categories. The labeled data then assists in pinpointing the corresponding positions of these categories in the medical records.

Our approach employs two sophisticated models for fine-tuning the task data: DistilBERT [10] and ELECTRA (Efficiently Learning an Encoder that Classifies Token Replacement Accurately) [11]. DistilBERT, a streamlined version of the BERT model, offers a balance between performance and efficiency. While maintaining a significant portion of BERT's capabilities, DistilBERT's reduced parameter count results in faster training and deployment. This model is particularly advantageous in scenarios requiring the processing of large datasets, where computational efficiency is key.

In contrast, ELECTRA introduces an adversarial training methodology, markedly improving the model's language comprehension and representation, especially in complex tasks. This model's resilience and enhanced performance, particularly in challenging contexts, are attributable to its unique training approach. Given the prevalence of medical terminology in our data, we have opted for a specialized variant, medical-ELECTRA [12]. This model, an adaptation of ELECTRA fine-tuned with a medical corpus, is tailored to comprehend the intricacies of medical records, thus ensuring the effective extraction of patients' private information.

The use of medical-ELECTRA, tailored specifically for medical terminology, is expected to significantly enhance the accuracy and reliability of SHI extraction in medical records.

2.3 Generative Approach

In addressing Task 1, we adopted a dual-model strategy, employing flan-t5-large [13] and Pythia-160m [14], as illustrated in Fig. 1. The flan-t5-large model, founded on a seq2seq

Table 1. Examples for the Voting Mechanism in Single and Dual Models.

Example	Prediction Type	Prediction Content	F-t5 Count	F-t5 Vote	Pythia Count	Pythia Vote	Dual Count	Dual Vote
1	DOCTOR	Ngarigo ARNSWOR	2	W	0		2	
1	DOCTOR	Ngarigo ARNSWORTH	1		4	R	5	R
2	DEPARTMENT	Oran	3	W	0		3	
2	DEPARTMENT	Orana	0		4	R	4	R
3	HOSPITAL	PORTLAND DISTRICT	2	W	0		2	
3	HOSPITAL	PORTLAND DISTRICT HEALTH	0		6	R	6	R
4	HOSPITAL	ST VINCENT'S HOSPITAL	1		0		1	
4	HOSPITAL	ST VINCENT'S HOSPITAL (LISMORE	2	W	0		2	
4	HOSPITAL	ST VINCENT'S HOSPITAL (LISMORE)	0		4	R	4	R
5	MEDICALRECORD	673633.LRE	2	R	1		3	R
5	MEDICALRECORD	673633.LRE CECCHI	0		1	W	1	
5	MEDICALRECORD	673633	0		1		1	
6	DOCTOR	I Klamn	4	R	1		5	R
6	DOCTOR	I Klamn, Validated	0		2	W	2	
7	HOSPITAL	HILLSTON DISTRICT HOSPITAL	3	R	0		3	R
7	HOSPITAL	SPECIMEN Received from the HILLSTON DISTRICT HOSPITAL	0		1	W	1	

(continued)

Table 1. (*continued*)

Example	Prediction Type	Prediction Content	F-t5 Count	F-t5 Vote	Pythia Count	Pythia Vote	Dual Count	Dual Vote
8	MEDICALRECORD	8511911.UJT	2	R	1		3	R
8	MEDICALRECORD	8511911.UJT Tidrington	0		1	W	1	
8	IDNUM	8511911	0		1		1	
9	PATIENT	Cartelli, Carter TEODORO	3	R	1		4	R
9	PATIENT	Cartelli	1		2	W	2	
9	PATIENT	Cartelli, Carter	0		1		1	
9	STREET	Carte	0		2		2	
10	HOSPITAL	ST GEORGE HOSPITAL	2	W	2		4	W
10	HOSPITAL	ST GEORGE HOSPITAL (QLD	0		1		1	
10	HOSPITAL	ST GEORGE HOSPITAL (QLD)	0		3	R	3	R

architecture with both transformer encoder and decoder structures [15], is adept at generating concise and focused text outputs. Conversely, Pythia-160m, operating as a causal language model with only a transformer encoder framework [16], specializes in producing more extended sequences. This dual-model combination offers a comprehensive and versatile approach to private data extraction.

Specifically, the flan-t5-large model was fine-tuned using the LoRA method [5], enhancing its efficiency in managing categories with sparse examples and improving its adaptability across diverse data types. For pythia-160m, we employed its 160 million parameter version, strategically freezing intermediate layers to optimize the training efficiency and minimize overfitting. The careful selection and specific tuning of flan-t5-large and Pythia-160m were crucial in meeting the demanding data processing needs of our task, ensuring both adaptability and efficiency.

For Task 2, we utilized the flan-t5-base model, fine-tuned for enhanced performance in time information normalization. This fine-tuning, conducted on a balanced dataset, aimed to improve the model's accuracy and minimize bias. Detailed parameters set for this process are outlined in Table 2.

Voting Mechanism In Task 1, a crucial aspect of our methodology was the implementation of a voting mechanism to integrate the outputs of both flan-t5-large and Pythia-160m models. This voting system was designed to evaluate and select the most accurate results from each model's predictions. By comparing the outputs, the mechanism enhanced the overall accuracy and reliability of our predictions. This approach was instrumental in

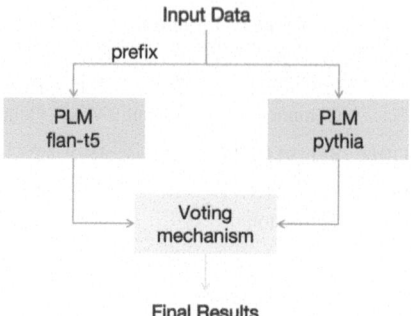

Fig. 1. Dual-model architecture utilized in the algorithm design for Task 1.

combining the strengths of each model, ensuring a more robust and precise identification of SHI elements in the medical records.

To elucidate the procedure of our dual-model voting mechanism, we analyzed several examples where the mechanism was applied using both single and dual-model configurations. As outlined in the "Data Grooming" section, our data preparation involved diverse formats, incorporating substantial redundant content, which led to varied model predictions for identical content segments. Through the voting mechanism, we synthesized these varied predictions to arrive at a conclusive outcome, detailed in Table 1. This table delineates the categories and content generated, alongside the voting outcomes from the flan-t5 and Pythia models, culminating in the dual-model's final output. Here, 'W' symbolizes an incorrect prediction, while 'R' signifies that both the content and type predictions were accurate. The insights reveal that exclusive reliance on the flan-t5 model may yield incomplete outputs, whereas the Pythia model alone tends to produce redundant outputs (in cases of tied votes, the longest answer is chosen). Merging the outputs from both models elevates the precision and recall rates of the predictions. Nevertheless, specific instances, such as Example 10, highlight scenarios where the dual-model approach results in inaccuracies. Despite occasional variances, our empirical analysis and statistical evaluations (as shown in Table 6) underscore the dual model's constructive influence on enhancing predictive accuracy.

Our training strategy for Task 1 was meticulously formulated to minimize overfitting risks and ensure stable convergence of the models. We utilized the AdamW optimizer [17], renowned for its effectiveness in controlling overfitting. A linear learning rate decay, accompanied by an initial warm-up phase, was employed to facilitate a well-balanced learning rate progression. This strategy helped in achieving a consistent and efficient model learning process. Furthermore, we implemented an early stopping protocol, halting the training when the validation error stopped decreasing, thus optimizing the model's performance while efficiently utilizing computational resources.

For data management, we adopted the use of task-specific prefixes for datasets from various sources, as detailed in Table 4. This approach was critical for maintaining a diverse and balanced training dataset, ensuring an appropriate mix of labeled and unlabeled data, and thereby preventing a bias towards non-informative outputs.

In Task 2, our focus was on adhering to strict time normalization protocols. The use of task-specific prefixes, as listed in Table 4, played a pivotal role in guiding the model towards generating outputs that align with the ISO 8601 standard. For the fine-tuning of this model, we utilized a randomly selected 15% subset of our dataset as the validation set, ensuring the model's robustness and effectiveness in normalizing time-related data.

Table 2. Parameters of Pretrained Models.

Task	Task 1	Task 1	Task 2
Batch size	16	16	64
Max length	128	196	32
Learning rate	1.00E-04	5.00E-05	1.00E-04
Weight decay	0.05	0.02	0.05
Pretrained Model	google/flan-t5-large	EleutherAI/pythia-160m	google/flan-t5-base
LoRA config	r = 16	\	\
	lora_alpha = 32		
	lora_dropout = 0.05		
Frozen layers	\	layer 2 - layer 9 (8 layers)	\
Null ratio	0.3	0.3	\

3   Experimental Setup

3.1   Data Processing

The OpenDeID corpus v2 dataset [18], derived from a collaboration with the Secure Research Environment for Digital Health (SREDH) Consortium [19] and sourced from the Lowy Cancer Research Centre at the University of New South Wales (UNSW) through the Health Science Alliance Biobank, is a meticulously de-identified collection of 3,244 electronic health records (EHRs) of surgical pathology reports [6]. Originally designed for research in automated information extraction from unstructured medical records, it serves as an essential resource for developing models to accurately identify and extract Sensitive Health Information (SHI) from textual documents. Characterized by its diversity in medical conditions and the length of reports, the dataset offers a broad foundation for analysis, with reports averaging around 600 words and some extending up to 2755 words. This variability affords a unique opportunity for model training, providing a wide spectrum of data that covers various aspects of medical documentation and patient information, thus ensuring a robust and inclusive approach to health informatics research.

The dataset comprises reports that vary significantly in length, providing a comprehensive basis for model development and evaluation. These datasets were instrumental in training and evaluating our proposed methods. Each dataset comprised multiple medical records, formatted as shown in Table the medical record example in Fig. 2a. Accompanying these datasets was an answer file for each medical record. This file detailed the

SHI by listing the report ID, SHI category, start and end position indices, and the actual SHI content, as exemplified in Table the answer example in Fig. 2a.

Table 3 presents the statistical overview of these datasets. On average, each document in the dataset contained over 600 words, with approximately 15 instances of structured health information per document. An important observation from our analysis was the coverage rate of SHI categories in each dataset, which stood at around 60%. This indicated a partial representation of SHI categories in any single dataset. To address this limitation, we adapted our data processing methodology to ensure comprehensive coverage and representation of various SHI types across the datasets.

Table 3. Dataset Statistics.

Item	First Phase Training	Second Phase Training	Validation
document counts	1120	614	560
avg. Line count per doc	59.76	51.21	57.73
max. Line count per doc	206	142	252
avg. Word count per doc	624.5	607.66	641.7
max. Word count per doc	2753	2084	2755
avg. Entity counts per doc	14	17.5	15.8
max. Entity counts per doc	42	42	53
avg. Entity coverage rate	0.25	0.31	0.28
entity coverage rate	0.6	0.6	0.28

Preparing Data for Discriminative Models: To address this task, considering the contextual information that might influence the effectiveness of SHI extraction, we design two methods to process data. The first one treats a single medical record as one data input. And the second one treats each line in a medical record as an individual input. We tokenize the input words based on space and annotate them using the IOB format as shown in the NER record example in Fig. 2b. In the processed data, each line represents a word followed by its SHI category, separated by a space. Additionally, each entry in the dataset is delineated by two newline characters.

Long Documents. Due to the input limitation of 512 tokens for the language model (PLM) and considering that the average word count of each pathological report is approximately 600 words, with the longest reaching up to 2755 words, it's not feasible to input an entire report directly. To account for contextual relevance, we have decided to extract 512 tokens every 64 characters within the document as one input. This approach ensures that each input captures some preceding context while also accommodating new textual content, facilitating a more comprehensive contextual understanding.

Preparing Data for the Generative Models in Task 1, our data processing involved grooming the data and augmenting categories with smaller sample sizes. For Task 2, we shifted our focus towards the extraction and normalization of task-specific data. This

Medical Records Example	Answer Example
Episode No: 09F016547J	10 IDNUM 14 24 09F016547J 10 MEDICALRECORD 25 35 091016.NMT
091016.NMT	10 PATIENT 37 50 SIZAR, HOWARD
	10 IDNUM 61 69 09F01654
SIZAR, HOWARD	10 STREET 70 77 Runford
Lab No: 09F01654	10 CITY 78 85 RENMARK
Runford	10 STATE 87 90 TAS
RENMARK TAS 5084	10 ZIP 92 96 5084
Specimen: Tissue	10 DATE 122 131 24/8/1993 1993-08-24
D.O.B: 24/8/1993	10 TIME 151 170 28/08/2013 at 08:26 2013-08-28T08:26
Sex: M	10 DEPARTMENT 182 192 St Vincent
Collected: 28/08/2013 at 08:26	10 HOSPITAL 193 229 BATLOW/ADELONG MULTI PURPOSE SERVICE
Location: St Vincent-BATLOW/ADELONG MULTI PURPOSE SERVICE	10 DOCTOR 233 249 JAXON AL-KARSTEN
DR JAXON AL-KARSTEN	
Distribution: FILE-COPY	
CLINICAL:	
Ca prostate. Prostate SV. G7 L base.	
MACROSCOPIC:	
...	

(a)

NER Dataset for Discriminative Models	Original Dataset for Generative Models
Episode O	10 1 Episode No: 09F016547J **IDNUM:09F016547J**
No: O	10 25 091016.NMT **MEDICALRECORD:091016.NMT**
90J789814B B-IDNUM	10 37 SIZAR, HOWARD **PATIENT:SIZAR, HOWARD**
903789.UDK B-MEDICALRECORD	10 52 Lab No: 09F01654 **IDNUM:09F01654**
HEAVILIN, B-PATIENT	10 70 Runford **STREET:Runford**
CANDACE I-PATIENT	
Lab O	Concatenated Dataset for Generative Models
90J78981 B-IDNUM	
Idonia B-STREET	10 1 Episode No: 09F016547J 091016.NMT SIZAR, HOWARD
TATURA B-CITY	Lab No: 09F01654 Runford RENMARK TAS 5084 Specimen:
2828 B-ZIP	Tissue **IDNUM:09F016547J++MEDICALRECORD:091016.NMT++**
Specimen: O	**PATIENT:SIZAR, HOWARD++IDNUM:09F01654++STREET:Runford**
Tissue O	**++CITY:RENMARK++STATE:TAS++ZIP:5084**
D.O.B: O	10 25 091016.NMT SIZAR, HOWARD Lab No: 09F01654 Runford
16/2/2020 B-DATE	RENMARK TAS 5084 Specimen: Tissue D.O.B: 24/8/1993 Sex: M
	MEDICALRECORD:091016.NMT++PATIENT:SIZAR, HOWARD++
	IDNUM:09F01654++STREET:Runford++CITY:RENMARK++STATE:TAS
	++ZIP:5084++DATE:24/8/1993

(b)

Fig. 2. Examples of the datasets used in the experiment. (a) presents the given medical data with the answer key. (b) shows data preparation for discriminative models (left), and examples of original and concatenated data inputs for generative models (right).

comprehensive approach was vital to optimize the effectiveness and accuracy of our generative models in both tasks.

Data Grooming. We customized our datasets for the generative model by creating three distinct forms of datasets:

- Original Dataset: This dataset focuses on preserving the integrity of the original data. We segmented the data based on newline characters to maintain its original structure. Each line in the dataset represents an individual instance, where the initial part is the model's input, and the subsequent bold part is the target output for the model to generate, as exemplified in the original dataset example for generative in Fig. 2b.
- Concatenated Dataset: To enrich the contextual information, we concatenated shorter entries from the original dataset, specifically those with less than 128 characters. This process not only augmented the contextual depth of the data but also included

overlapping segments to enhance the model's recall and precision. Figure 2b showcases two instances from this dataset. In each instance, the initial content represents the concatenated input for the model, while the subsequent bold part, connected by "++", is the model's generation target. This approach was designed to challenge the model to recognize and generate coherent outputs from complex, concatenated input sequences.

- Truncated Dataset: For entries exceeding 128 characters, we employed truncation using a 30-character sliding window for efficiency. This approach allowed us to create a dataset optimized for both resource usage and training.

The introduction of overlapping and redundant entries was an inherent byproduct of these processes. To address this and ensure fair evaluation, we generated distinct training and test datasets from unique file IDs, with the test set comprising approximately 15% of the total samples. The choice of a 15% test set size is based on standard practices in machine learning and deep learning, aiming to provide a balanced dataset for both training and evaluation while ensuring the model's generalizability to new, unseen data. This proportion is widely accepted across various datasets and applications, including healthcare [20].

Furthermore, we assigned specific prefixes to each dataset category for the flan-t5 model training, as outlined in Table 4. This strategy significantly enhanced the model's ability to understand and process various types of data, ensuring more accurate and reliable outputs.

Small Sample Augmentation. To strengthen the model's effectiveness in categories with smaller sample sizes, we employed augmentation techniques:

- Target Content Replacement: Through regular expression searches, we identified and replaced target content within sentences to generate new training examples, thereby increasing the diversity of instances without altering sentence structure.
- Random Noise Insertion and Deletion: We introduced random noise and deleted non-target words in approximately 15% of the sentences, aiming to improve the model's resilience and adaptability to data variations in small sample scenarios.

These augmentation strategies specifically targeted the categories with smaller sample sizes. By diversifying the data in this manner, we aimed to improve the model's ability to accurately recognize and process fewer common data categories, thereby enhancing overall performance in handling rare data instances.

Normalization Data Extraction In Task 2, our approach leveraged the data extracted from Task 1 to fine-tune models for time information normalization. This process involved handling data in both its pre-normalization (unstandardized) and post-normalization (standardized) states. Specifically, we paired each category with its corresponding unstandardized data as the input, and the model's task was to produce a normalized, standardized output. This methodology was designed to enable the model to learn the transformation of time-related data from its original form to a standardized format compliant with the ISO 8601 standard. Table 5 provides examples from this normalization dataset, showcasing how the input data is transformed into its normalized

Table 4. Task Prefixes.

Task	Dataset	Task Prefix
Task 1	Original data	Private information extraction from patients' records:
	Sliced data	Private information extraction from sliced patients' records:
	Spliced data	Private information extraction from spliced patients' records:
Task 2	Normalization	Time information regularization according to ISO 8601 standard:

Table 5. Normalization Dataset Examples.

Input	Output
DATE:22/12/2005	2005/12/22
DURATION:26-29 yrs	P27.5Y
TIME:2683-04-13 00:00:00	2683-04-13T00:00:00
DATE:5/3/63	2063/5/3

equivalent. This visual representation aids in understanding the practical application of the model's normalization capabilities.

Furthermore, during the fine-tuning phase of the Flan-t5-base model, we utilized specific task prefixes to guide the model more effectively. These prefixes, outlined in Table 4, were instrumental in directing the model's focus towards the desired normalization standards. By embedding these prefixes, we aimed to enhance the model's accuracy in interpreting and standardizing time-related data, ensuring consistency and reliability in the normalization process.

3.2 Evaluation Metrics

Our evaluation metrics for the competition were based on three critical attributes for each SHI instance: the starting position, the ending position, and the associated category in the text. We defined our evaluation criteria as follows:

- True Positive (TP): Correct identification of both the start and end positions, as well as the SHI category.
- False Positive (FP): Any mismatch in the start/end position or the SHI category identified by our method compared to the annotated SHI.
- False Negative (FN): Failure to identify a SHI instance that matches an annotated one in the text.

We specifically chose precision, recall, and F1-score, incorporating both micro and macro averages, due to their proven effectiveness in performance evaluation. These metrics crucially balance accuracy (precision) and comprehensiveness (recall) of entity identification, with the F1-score providing a unified measure of overall model performance. The use of micro and macro averages allows us to detail analyze model performance across diverse entity classes, addressing inherent class imbalances in NER tasks. Micro-averaging offers insight into the model's instance-level performance, while macro-averaging illuminates its ability to recognize less frequent entities, ensuring a holistic evaluation. These metrics are supported by literature that evidences the application and success of these metrics in related research such as in healthcare NER task [21] and cross-domain NER recognition [22]. And we believe that these metrics directly address our primary concerns of accuracy and completeness in the SHI extraction and time normalization tasks.

3.3 Environments

The experimental setup was primarily conducted on Google Colab, utilizing its virtual machine environment and GPU resources, including NVIDIA Tesla V100 (16 GB) and A100 (40 GB). The primary development language for our experiments was Python, version 3.8. For our main algorithmic framework, we employed PyTorch.

4 Experiment Results

Our team, Team 4593, demonstrated strong performance in the competition, achieving a metric of 0.7668 in the SHI extraction task (Task 1), ranking us 12th. In the Time Information Normalization task (Task 2), our improved methodology secured us 7th place with a performance metric of 0.7959. Overall, we were ranked 9th, underscoring our methodologies' success in tackling privacy protection and medical data standardization challenges.

4.1 Experiment Results for SHI Extraction Task

Our evaluation for Task 1 demonstrated the superiority of integrated model configurations over standalone models. The data, as shown in Table 6, reveals that setups combining flan-t5 and Pythia models significantly outperformed others in terms of F1 scores, especially in the Macro F1 metric. The discriminative model SentNER-DistilBERT emerged as the top performer within its category. However, the overall low Micro F1 scores (below 0.1) for discriminative models indicated a potential misalignment with the specific requirements of the task. This finding suggests a need for deeper analysis to understand the limitations of discriminative models in this context.

In contrast, the generative models, and notably the combination of flan-t5 and Pythia, exhibited remarkable effectiveness. This is visually represented in Fig. 3, where the combined model showed superior performance in terms of recall and Macro F1 metrics. Notably, the integration of flan-t5-large (using LoRA) with Pythia-160m (Frozen) displayed high precision across various data categories. This is detailed in Table 7, which presents the performance metrics from the public leaderboard.

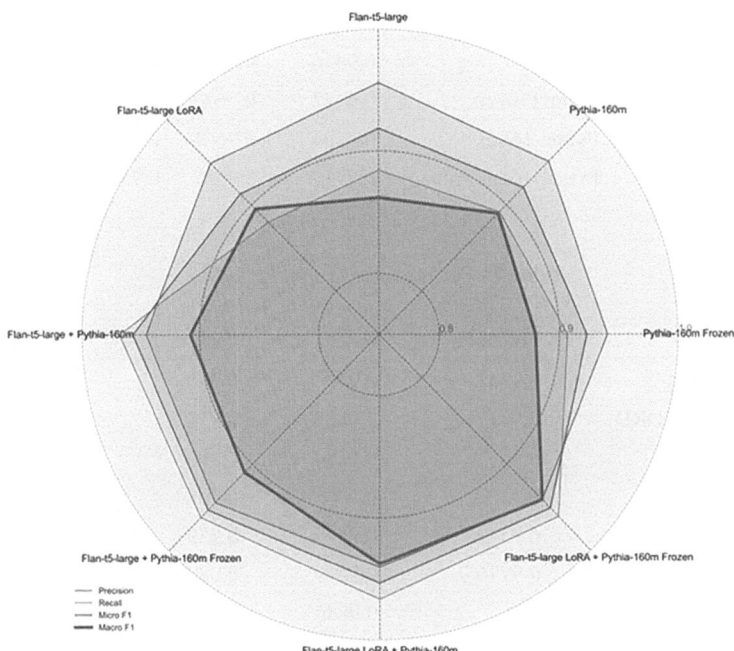

Fig. 3. Performance of models and model combinations in radar chart.

Table 6. Performance of Different Pretrained Models and Model Combinations.

Model Configuration	Precision	Recall	Micro F1	Macro F1
Discriminative model				
DocNER-DistilBERT	0.0031	0.0082	0.0152	-
DocNER-ELECTRA	0.0052	0.0102	0.0174	-
DocNER-medicalELECTRA	0.0034	0.0057	0.014	-
SentNER-DiltilBERT	0.0633	0.0234	0.0459	-
SentNER-medicalELECTRA	0	0	0	-
Generative model				
Pythia-160m Frozen	0.941	0.9068	0.9236	0.88
Pythia-160m	0.9498	0.8919	0.9199	0.8901
Flan-t5-large	0.956	0.8842	0.9187	0.8619
Flan-t5-large LoRA	0.9483	0.8802	0.913	0.8955

(continued)

Table 6. (*continued*)

Model Configuration	Precision	Recall	Micro F1	Macro F1
Flan-t5-large + Pythia-160m	0.9456	0.9682	0.9568	0.9077
Flan-t5-large + Pythia-160m Frozen	0.9442	0.9613	0.9527	0.9092
Flan-t5-large LoRA + Pythia-160m	0.9404	0.9665	0.9533	0.9375
Flan-t5-large LoRA + Pythia-160m Frozen	0.9413	0.9607	0.9509	0.9413

Table 7. Performance Details of Task 1.

Coding Type	Precision	Recall	F-measure	Support
IDNUM	0.9841	0.949	0.9663	1177
MEDICALRECORD	0.9617	0.9805	0.971	563
PATIENT	0.88	0.9284	0.9036	545
STREET	0.9122	0.9065	0.9094	321
CITY	0.9379	0.9407	0.9393	337
STATE	0.9776	0.9839	0.9807	310
ZIP	1	0.9846	0.9922	325
DATE	0.9778	0.9802	0.979	1617
DEPARTMENT	0.963	0.9235	0.9428	366
HOSPITAL	0.9542	0.9494	0.9518	593
DOCTOR	0.8836	0.8279	0.8549	2191
TIME	0.9701	0.9262	0.9476	420
AGE	0.8413	0.9298	0.8833	57
ORGANIZATION	0.6	0.625	0.6122	24
DURATION	0.8571	1	0.9231	6
PHONE	1	1	1	2
COUNTRY	1	1	1	2
LOCATION-OTHER	0.6667	0.5	0.5714	4
Micro-avg. F	0.9409	0.9229	0.9318	8860
Macro-avg. F	0.9093	0.9075	0.9084	8860

Another observation in Table 6 is that categories with a larger number of test samples, such as DATE, MEDICALRECORD, and HOSPITAL, achieved higher and more consistent results. This contrasts with categories with smaller sample sizes like ORGANIZATION and LOCATION-OTHER, which posed greater challenges. The performance in these categories was heavily reliant on specialized data processing techniques, significantly affecting the overall Macro F1 score. This underscores the crucial role of

Table 8. Performance Details of Task 2.

Temporal Type	Precision	Recall	F-measure	Support
DATE	0.8271	0.8108	0.8189	1617
TIME	0.9614	0.8905	0.9246	420
DURATION	1	1	1	6
Micro-avg	0.854	0.8277	0.8407	2043
Macro-avg	0.9295	0.9004	0.9147	2043

sample size in determining model performance across different categories, highlighting the importance of robust data processing techniques in managing categories with fewer samples.

4.2   Experiment Results for Time Information Normalization Task

In the TIN Task, the flan-t5-base model demonstrated impressive performance. This effectiveness is substantiated by its handling of diverse temporal expressions, with results detailed in Table 8 from the public leaderboard. Notably, the model showed a capability in processing small sample categories like DURATION, which comprised only 6 data points. This achievement can be primarily credited to our targeted data augmentation techniques.

Moreover, the model displayed consistent performance across categories with larger sample sizes, paralleling the stability seen in Task 1. The well-balanced performance across different time normalization categories, as presented in Table 8, underscores the model's comprehensive effectiveness. It highlights the model's adaptability in dealing with both data-rich and data-sparse categories, demonstrating its all-around efficacy in normalizing temporal information.

5   Discussion

5.1   Generative Models vs. Discriminative Models for the SHI Extraction Task

Our experiments across different models and configurations have led to some intriguing insights. Initially, we hypothesized that discriminative models would be ideally suited for the SHI extraction task, which is the NER task, based on their conventional application in similar contexts. However, the results depicted in Table 6 suggest an alternative narrative. We observed that generative models outperformed discriminative models across almost all evaluated categories. This surprising outcome highlights the formidable capability of generative models in managing complex and varied data types, as demonstrated by their superior precision, recall, and F-measure performance metrics.

The superior performance of generative models can be attributed to their adaptability in managing complex NER tasks, which often involve diverse categories and precise position matching. The generative models' proficiency in generating coherent text and predicting context, leveraging LLMs, played a critical role. Additionally, the intricacies involved in data post-processing, particularly when dealing with the IOB tagging format, might have presented substantial challenges to discriminative models, more so in the context of data augmentation. The varied nature of the data and the complexities associated with augmenting samples in categories with uneven distribution also likely contributed to the diminished performance of discriminative models.

Case 1: Preventing Overly Long Outputs

"('1057', '7761')": [
    {
        "count": 9,
        "output": "1057\tHOSPITAL\t7714\t7761\tEASTERN HEALTH BLACKBURN PUBLIC SURGICAL CENTRE"
    },
    {
        "count": 1,
        "output": "1057\tDEPARTMENT\t7714\t7728\tEASTERN HEALTH"
    },
    {
        "count": 3,
        "output": "1057\tHOSPITAL\t7663\t7761\tSUPPLEMENTARY REPORT: 13.3.63 Consultation/Review EASTERN HEALTH BLACKBURN PUBLIC SURGICAL CENTRE"
    }
],

Case 2: Preventing Incomplete Outputs

"('1002', '36')": [
    {
        "count": 5,
        "output": "1002\tMEDICALRECORD\t25\t36\t4672273.RKT"
    },
    {
        "count": 2,
        "output": "1002\tMEDICALRECORD\t25\t31\t467227"
    }
],
"('1002', '54')": [
    {
        "count": 7,
        "output": "1002\tPATIENT\t38\t54\tParmar, Gregorio"
    },
    {
        "count": 1,
        "output": "1002\tPATIENT\t38\t44\tParmar"
    },
    {
        "count": 1,
        "output": "1002\tDOCTOR\t46\t54\tGregorio"
    }
],

Case 3: Avoiding Incorrect Categorization

"('1011', '98')": [
    {
        "count": 1,
        "output": "1011\tPATIENT\t91\t98\tFranich"
    },
    {
        "count": 10,
        "output": "1011\tSTREET\t91\t98\tFranich"
    }
],

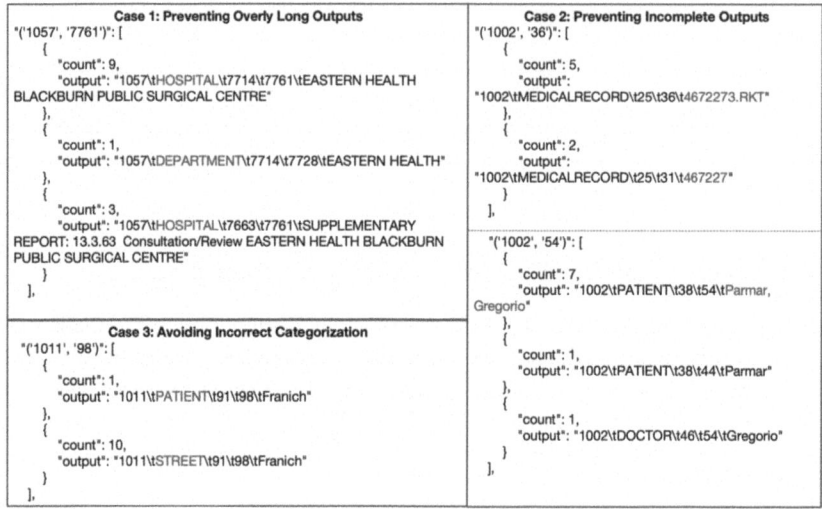

Fig. 4. Successful cases in dual model with voting mechanism for SHI extraction.

Case 1: Failed to identify the most complete target content

"('1001', '63')": [
    {
        "count": 3,
        "output": "1001\tPATIENT\t37\t63\tVatterott, Jerrie CLARENCE"
    },
    {
        "count": 1,
        "output": "1001\tPATIENT\t37\t46\tVatterott"
    },
    {
        "count": 5,
        "output": "1001\tPATIENT\t37\t54\tVatterott, Jerrie"
    }
],

Case 2: Incorrect category determination

"('1026', '218')": [
    {
        "count": 4,
        "output": "1026\tDEPARTMENT\t195\t218\tPRINCE OF WALES PRIVATE"
    },
    {
        "count": 1,
        "output": "1026\tDEPARTMENT\t195\t213\tPRINCE OF WALES PR"
    },
    {
        "count": 5,
        "output": "1026\tHOSPITAL\t195\t218\tPRINCE OF WALES PRIVATE"
    }
],

Fig. 5. Failure cases in dual model with voting mechanism for SHI extraction.

These observations lead us to reconsider the traditional model selection strategies for NER tasks. The compelling performance of generative models, even in categories with smaller sample sizes, coupled with their consistency in larger sample categories, underscores the effectiveness of nuanced data processing techniques. It suggests that for NER tasks, a comprehensive evaluation of both the task characteristics and data processing methodologies is crucial.

5.2 Voting Mechanism for the SHI Extraction Task with Dual-Model Architecture

In the SHI extraction task, we employed a novel voting mechanism that capitalized on the complementary strengths of auto-regressive models Pythia, and auto-encoder models flan-t5. Pythia models tend towards generating lengthier content, while flan-t5 models are inclined to produce shorter outputs. Our voting mechanism, applied to various forms of the dataset (original, concatenated, and truncated), significantly enhanced the decision-making efficacy of the combined model. This approach effectively harnessed the generative abilities of both models to strike a balance between output length and completeness.

The advantages of this voting mechanism are multifaceted. Primarily, it mitigates the issue of excessively long outputs, a common trait in pythia model outputs, as exemplified in case 1 in Fig. 4. Similarly, it addresses the problem of incomplete outputs, often seen with the flan-t5 model, as illustrated in case 2 of the same figure. Furthermore, it plays a crucial role in reducing incorrect categorizations. However, the mechanism is not without its challenges, particularly when the voting results are neck and neck. This can lead to suboptimal decisions in content generation or category classification, as depicted in Fig. 5. Instances where the models favor potentially optimal solutions over technically correct but suboptimal ones, have highlighted areas for further improvement. These instances underscore the need for continuous refinement in both the training methodologies and the criteria used for judgment within the voting mechanism.

5.3 Model Performance Analysis in Time Normalization Task

In the time normalization task, our findings indicate that using the same model for both SHI extraction and time normalization often led to suboptimal results, particularly in time normalization. This disparity in performance is primarily due to the intrinsic differences between these tasks. SHI extraction involves directly deriving content from the model's input, whereas time normalization demands the transformation of content into a format not explicitly present in the input.

A significant insight from our study was the critical role of clear, task-specific prefixes in achieving successful time normalization. By incorporating a specific normalization standard into the model's prefix, particularly one aligning with the "ISO 8601" format, we were able to notably improve the flan-t5 base model's accuracy in standardizing time-related data. This approach highlights the importance of providing the model with precise guidance and standards. Such targeted directives were key in enhancing the model's ability to accurately perform time normalization tasks, demonstrating the efficacy of clear and explicit instructions in guiding the model's processing capabilities.

6 Conclusion

Our study ventured into the use of discriminative and generative models to enhance privacy protection and standardize medical data. We concentrated on efficiently identifying and standardizing SHI using pre-trained LLMs. Contrary to initial expectations, generative models surpassed traditional discriminative models in the SHI extraction task.

We further refined this approach by implementing a dual-model fusion coupled with a voting mechanism, leading to significant improvements in the precision, recall, and F1 scores in the SHI extraction task.

Key innovations of our research include:

- Diverse Data Forms and Small Sample Augmentation: Our method incorporated various data forms for model input and enhanced small sample data. This approach provided a more comprehensive training dataset and bolstered the model's ability to handle a range of data types and scenarios, particularly beneficial for underrepresented categories.
- Dual-Model Fusion and Voting Mechanism: The combination of flan-t5 and Pythia models, exploiting their unique strengths, was optimized through a voting mechanism. This strategy effectively addressed the biases typically present in generative models, resulting in more precise data processing.

These strategies significantly elevated the model's performance in data standardization tasks, marking a step forward in the field of medical data annotation. Nonetheless, our study encountered certain limitations and challenges, such as difficulties in specific data categories and model biases, which could impact the reliability and validity of our results. Future research directions we aim to pursue include:

- Finer-grained Loss Functions: The development of custom loss functions tailored to optimize macro F1 values in small sample categories.
- Advanced Model Fusion Techniques: We plan to delve into more intricate methods for integrating model outputs, such as developing linear layers that merge the strengths of both discriminative and generative models, thus deepening the analysis and enhancing accuracy.
- PEFT Techniques' Expanded Application: We intend to further explore Parameter-efficient Fine-tuning (PEFT) techniques like adapters and prompt tuning to improve overall model performance.

Our research contributes significantly to medical data processing and privacy protection, showcasing the potential of deep learning techniques in these domains. By providing effective solutions for data privacy and quality, our work offers valuable insights and references for AI applications in healthcare. We hope this study inspires further in-depth exploration and innovation in this crucial field.

Acknowledgments. We extend our deepest gratitude to the Secure Research Environment for Digital Health (SREDH) Consortium for their generous provision of data and support. Additionally, we would like to express our sincere appreciation to the organisers of AICup2023, including the SREDH Consortium, National Kaohsiung University of Science and Technology, and Asian University. Their dedication and hard work in organizing this event have provided a unique platform for sharing knowledge, fostering collaboration, and showcasing the latest advancements in digital health research. To all individuals and organizations from the partner entities who have joined the consortium and contributed to the SREDH Platform, your expertise and support have been invaluable. We are grateful for the opportunity to be a part of this vibrant community and look forward to future collaborations.

Disclosure of Interests.   The authors have no competing interests to declare that are relevant to the content of this article.

References

1. AI-Cup 2023: Privacy Protection and Medical Data Standardization Challenge (2023). https://www.sredhconsortium.org/sredh-competitions/sredhai-cup-2023. Accessed 03 Apr 2024
2. Yuan, J., Tang, R., Jiang, X., Hu, X.: Large language models for healthcare data augmentation: an example on patient-trial matching. In: AMIA Annual Symposium Proceedings, vol. 2023, p. 1324. American Medical Informatics Association (2023)
3. Durango, M.C., Torres-Silva, E.A., Orozco-Duque, A.: Named entity recognition in electronic health records: a methodological review. Healthc. Inform. Res. **29**, 286 (2023)
4. Hu, E.J., Shen, Y., Wallis, P., Allen-Zhu, Z., Li, Y., Wang, S., et al.: Lora: low-rank adaptation of large language models, arXiv preprint arXiv:2106.09685 (2021)
5. Codalab competition (2024). https://codalab.lisn.upsaclay.fr/competitions/15425?secret_key=db7687a5-8fc7-4323-a94f-2cca2ac04d39#learn_the_details-overview. Accessed 03 May 2024
6. Huang, P., Liu, T.-E.: Resources for AICUP competition on deidentification of electronic medical record notes (2024). https://github.com/paveenH/AIcup. Accessed 03 May 2024
7. Rasmy, L., Xiang, Y., Xie, Z., Tao, C., Zhi, D.: Med-BERT: pretrained contextualized embeddings on large-scale structured electronic health records for disease prediction. NPJ Digit. Med. **4**, 86 (2021)
8. ISO 8601 - Simple English Wikipedia, the free encyclopedia. https://simple.wikipedia.org/wiki/ISO_8601. Accessed 03 May 2024
9. Sanh, V., Debut, L., Chaumond, J., Wolf, T.: Distilbert, a distilled version of BERT: smaller, faster, cheaper and lighter. arXiv:1910.01108 (2020)
10. Clark, K., Minh-Thang Luong, Q.V.L., Manning, C.D.: Electra: pre-training text encoders as discriminators rather than generators. In: Proceedings of the International Conference on Learning Representations (2020)
11. fspanda, electra-medical-small-discriminator (2020). https://huggingface.co/fspanda/electra-medical-small-discriminator. [Online; huggingface]
12. Chung, H.W., et al.: Scaling instruction-finetuned language models, arXiv preprint arXiv:2210.11416 (2022)
13. Biderman, S., et al.: Pythia: a suite for analyzing large language models across training and scaling. In: International Conference on Machine Learning, pp. 2397–2430. PMLR (2023)
14. Vaswani, A., et al.: Attention is all you need. In: Advances in Neural Information Processing Systems, vol. 30 (2017)
15. Black, S., et al.: GPT-neox-20b: an open-source autoregressive language model, arXiv preprint arXiv:2204.06745 (2022)
16. Loshchilov, I., Hutter, F.: Decoupled weight decay regularization. In: Proceedings of the International Conference on Learning Representations (2019)
17. Jonnagaddala, J., Chen, A., Batongbacal, S., Nekkantti, C.: The opendeid corpus for patient de-identification. Sci. Rep. **11**, 19973 (2021). https://doi.org/10.1038/s41598-021-99554-9
18. Secure Research Environment for Digital Health (SREDH) Consortium (2024). https://www.sredhconsortium.org/. 03 May Accessed
19. Kumaar, M.A., Samiayya, D., Vincent, P.D.R., Srinivasan, K., Chang, C.-Y., Ganesh, H.: A hybrid framework for intrusion detection in healthcare systems using deep learning. Front. Publ. Health **9** (2021)
20. Tarcar, A.K., Tiwari, A., Dhaimodker, V.N., Rebelo, P., Desai, R., Rao, D.: Healthcare NER models using language model pretraining, arXiv preprint arXiv:1910.11241 (2019)

21. Liu, Z., et al.: CrossNER: evaluating cross-domain named entity recognition. In: Proceedings of the AAAI Conference on Artificial Intelligence, vol. 35, pp. 13452–13460 (2021)
22. Liu, J., et al.: OpenDeID pipeline for unstructured electronic health record text notes based on rules and transformers: deidentification algorithm development and validation study. J. Med. Internet Res. [Internet]. **25**(1), e48145 (2023). https://www.jmir.org/2023/1/e48145
23. Chen, A., Jonnagaddala, J., Nekkantti, C., Liaw, S.T.: Generation of surrogates for de-identification of electronic health records [Internet]. ebooks.iospress.nl, pp. 70–3. IOS Press (2019). https://ebooks.iospress.nl/publication/51950. Accessed 19 Apr 2024
24. Alla, N.L.V., et al.: Cohort selection for construction of a clinical natural language processing corpus. Comput. Methods Programs Biomed. Update **1**, 100024 (2021)
25. Mir, T.H., et al.: Deidentification and temporal normalisation of the electronic health record notes using large language models: the SREDH/AI-Cup 2023 deidentification competition. In: 2024 International workshop on deidentification of electronic medical record notes. Springer, Kaohsiung (2024)

Evaluation of OpenDeID Pipeline in the 2023 SREDH/AI-Cup Competition for Deidentification of Sensitive Health Information

Shalini Gupta[1] , Naga Lalitha Valli Alla[2] , Omkar Panchal[1] , Jan Witowski[3] ,
and Jitendra Jonnagaddala[4,5]([✉])

[1] CGD Health Pvt. Ltd., Hyderabad, Telangana, India
{Shalini,Omkar}@cgdhealth.com
[2] Royal Flying Doctor Service Western Operations, Perth, Australia
Lalitha.alla@rfdswa.com.au
[3] Ataraxis AI, New York, USA
jan.witowski@ataraxis.ai
[4] School of Population Health, UNSW Sydney, Kensington, Australia
z3339253@unsw.edu.au
[5] NMC Royal Hospital, Khalifa City, Abu Dhabi, United Arab Emirates

Abstract. The transformative influence of Electronic Health Records (EHRs) is manifested through their facilitation of streamlined management and dissemination of health data among healthcare practitioners. These fosters enhanced collaboration in patient care and contribute to superior healthcare outcomes. This phenomenon within healthcare informatics underscores the integration of EHRs with sophisticated technologies such as artificial intelligence and machine learning. The intricate functionalities of EHR are essential for precision, efficiency, and continuity in patient care. Deidentification is a crucial process for preserving patient privacy, facilitating safe data analysis, promoting public trust, and enabling global collaboration, especially in healthcare to accelerate research and response to public health crises. By anonymizing personal information, deidentification not only mitigates the risk of privacy breaches but also promotes data sharing and collaboration in medical research, advancing scientific discovery while respecting individual privacy rights. On the other hand, it poses challenges in balancing data utility and protection. In response to these challenges, the AI Cup 2023 - Privacy Protection and Medical Data Standardization Competition and the 2024 International Workshop on Deidentification of Electronic Medical Record Notes (IW-DMRN) were organized. We participate in AI Cup 2023 competition by employing a hybrid approach to deidentification across diverse datasets, aiming to enhance and refine techniques for safeguarding privacy in healthcare data utilization. The research contributes valuable insights for advancing comprehension and strategies in protecting sensitive information within varied datasets. The study utilized two separate corpora: the 2014 i2b2/UTHealth deidentification corpus and the 2016 CEGS N-GRID deidentification corpus, in addition to the OpenDeID v2 corpus provided in the competition. These datasets underwent processing and pre-processing stages, encompassing various steps. The deidentification method employed the OpenDeID pipeline which involved segmentation,

J. Jonnagaddala et al. (Eds.): IW-DMRN 2024, CCIS 2148, pp. 107–119, 2025.
https://doi.org/10.1007/978-981-97-7966-6_8

tokenization, tagging, and labeling, following a structured BIESO tagging scheme. The experimental setup incorporated a BERT-base model, fine-tuned on discharge summaries. The output underwent evaluation by the competition organizer using their evaluation metrics encompassed micro and macro averaged precision, recall, and F1-scores. The model demonstrated superior performance in Run1, achieving macro-averaged F1-scores of 0.7596718, 0.6969127, and 0.7269402 for precision, recall, and F-measure, respectively. The model highlighted high accuracy in deidentifying electronic health record text notes. The evaluation results affirm the model's efficiency and reliability in precise deidentification tasks. The results provide significant insights into the ethical and privacy-conscious utilization of healthcare data within dynamic multicenter healthcare frameworks.

Keywords: Deidentification · EHR text notes · Sensitive health information · Multicenter Corpora · BioBERT · OpenDeID

1   Introduction

Electronic Health Records (EHRs) represent a paradigm shift in healthcare informatics, transcending traditional paper-based medical documentation systems [1]. Beyond a mere digitization of medical records, EHRs embody a sophisticated integration of technologies, encompassing databases, standardized coding systems, and advanced software algorithms [2]. This convergence facilitates the seamless exchange of patient data among diverse healthcare entities, fostering a holistic and real-time view of a patient's medical history [3]. EHRs not only serve as repositories for clinical data but also function as dynamic platforms that support clinical decision-making through the integration of artificial intelligence and machine learning. As healthcare systems globally transition towards a more interconnected and data-driven future, the intricate functionalities of EHRs emerge as linchpins in ensuring precision, efficiency, and continuity in patient care. Machine Learning (ML), Artificial Intelligence (AI) and Large Language Models (LLMs) significantly enhance EHR by introducing advanced capabilities that revolutionize healthcare practices [4, 5]. These technologies contribute to the optimization of clinical workflows through automation, reducing the administrative burden on healthcare professionals. ML algorithms analyze vast datasets to provide predictive analytics, aiding in the early detection of potential health issues and enabling personalized treatment plans [6]. AI-powered Clinical Decision Support Systems (CDSS) offer valuable insights, improving the accuracy of diagnoses and treatment decisions [7]. Natural Language Processing (NLP) enhances the extraction of meaningful information from unstructured clinical notes, contributing to more comprehensive and context-aware patient records [8]. Furthermore, the continuous learning nature of ML ensures that EHR systems evolve with new medical knowledge, providing up-to-date and effective support for healthcare professionals in delivering high-quality patient care. Deidentification, an intricate process involving the anonymization of sensitive patient data within EHR text notes, is imperative in preserving patient privacy [9, 10]. This harmonious integration of EHR text notes and deidentification strategies not only underscores the technological sophistication required in contemporary healthcare systems but also establishes a robust framework for ethically sound and privacy-respecting data utilization [11]. Deidentifying EHR text notes presents several challenges, primarily stemming from the need to balance data utility with patient privacy. Striking an optimal equilibrium between rendering data anonymous and maintaining its usefulness for research and analysis is a complex task. Challenges include the potential loss of clinical context during deidentification, the variability in the effectiveness of deidentification methods across diverse types of data, and the risk of reidentification through the combination of innocuous information. Furthermore, ensuring the consistency and accuracy of deidentification processes across diverse datasets and evolving healthcare information structures adds an additional layer of complexity [12]. Nevertheless, existing deidentification methodologies encounter various limitations. These techniques exhibit inadequacies in their robustness and necessitate extensive rule modifications to enhance accuracy, especially when dealing with EHR text notes from diverse healthcare environments. Previous deidentification investigations primarily revolved around either rule-based strategies or hybrid methodologies integrating Conditional Random Fields (CRF) or Bidirectional Long Short-Term Memory (Bi-LSTM) models along with rule systems to detect sensitive health information (SHI) categories [13, 14]. Limited research has delved into exploring the potential of hybrid methods that amalgamate rules, innovative pretrained language models, and deep learning techniques.

To address these challenges AI Cup 2023- Privacy Protection and Medical Data Standardization competition [15, 16] and the 2024 International Workshop on the deidentification of Electronic Medical Record Notes (IW-DMRN) [17] were organized, and it was crucial for fostering trust in the deidentification process and promoting the responsible and ethical use of healthcare data for research and analysis. The deidentification of EHR text notes emerges as a vital component in guaranteeing the confidentiality of patient information and promoting responsible and ethical usage of healthcare data within diverse datasets. AI Cup 2023 competition presents a platform for participants to tackle real-world issues surrounding privacy protection and data standardization in medical records. By providing de-identified EHR data from authentic medical institutions, the competition challenges participants to address two sub-competition tasks: retrieving patient privacy information and normalizing time information. In the first subtask, participants are challenged to develop models capable of accurately identifying and extracting sensitive health information (SHI) such as patient names, addresses, dates of birth, and other SHI from EHR text notes. This challenge lies in devising robust algorithms that can effectively discern and extract such sensitive data while ensuring the preservation of patient privacy and data integrity. The second subtask focuses on the normalization of time information within EHR text notes. Participants are tasked with developing algorithms capable of standardizing timestamps and temporal references present in EHR text notes. This involves converting diverse time formats and expressions into a unified, standardized format, facilitating consistent analysis and interpretation of temporal data across different medical records.

Our participation in the competition aimed to employ a deidentification using a hybrid-based methodology called OpenDeID pipeline [18] to effectively extract SHI from EHR text notes. OpenDeID adopts a step-by-step pipeline strategy, integrating associative rules, supervised deep learning, and pretrained language models, to effectively deidentify sensitive information. Leveraging multicenter corpora, our objective was to develop and implement an advanced approach that integrates diverse datasets from multiple healthcare institutions [19]. This hybrid approach aimed to enhance the accuracy and efficiency of SHI extraction while ensuring the preservation of patient privacy across a wide spectrum of healthcare contexts.

2 Materials and Methods

2.1 Datasets

The dataset allocated for AI Cup 2023 is the OpenDeID v2 corpus of 3,244 reports. Specifically, this corpus comprises 2,100 pathology reports sourced from 1,833 distinct cancer patients across four urban hospitals in Australia. Within this corpus, there are a total of 38,414 instances of SHI. The construction of this corpus involved collaboration between two annotators in three separate experimental settings from the OpenDeID v1 corpus [20, 21]. Additionally, an additional 1,144 reports were extracted from the same four urban hospitals in Australia. OpenDeID v2 corpus is available through the Secure Research Environment for Digital Health (SREDH) Consortium [22].

In our study, we also utilized two additional corpora: the 2014 i2b2/UTHealth deidentification corpus and the 2016 CEGS N-GRID deidentification corpus. The 2014 i2b2/UTHealth deidentification corpus comprised a total of 1,304 longitudinal clinical narratives from 296 patients in the USA. Within this corpus, 28,872 SHI instances were annotated and categorized into six SHI categories and 25 subcategories [23, 24]. Another corpus utilized was the 2016 CEGS N-GRID deidentification corpus, consisting of 1,000 psychiatric notes from the USA [25]. Access to the 2014 i2b2/UTHealth deidentification and 2016 CEGS N-GRID deidentification datasets is facilitated through the official i2b2 website [26]. Interested parties can acquire the data by completing and submitting a Data Use Agreement (DUA) form, ensuring adherence to ethical and legal considerations associated with dataset utilization. The OpenDeID v2 corpus provided for AI Cup 2023 was partitioned into four subsets: the first part for training (1120 OpenDeID v2 corpus), the second part for further training (614 OpenDeID v2 corpus), the validation sets (OpenDeID v2 corpus) and the test sets (OpenDeID v2 corpus). Additionally, to enhance the training process and evaluate the OpenDeID pipeline comprehensively, the 2014 i2b2 and 2016 CEGS N-GRID datasets were incorporated into the training corpus. This inclusion allows for a more robust assessment of the OpenDeID pipeline's performance on a broader range of data, beyond the provided training datasets. The training set served for the initial model training, while the validation sets were utilized to fine-tune hyperparameters and implement early stopping. Subsequently, the models underwent training using the amalgamation of the training and validation sets, and their performance was evaluated on the designated test set. Table 1 illustrates the distribution of datasets across different runs. It outlines the breakdown of datasets for training, validating, and testing purposes. The table provides an overview of the datasets used in each run, including the number of reports.

Table 1. Distribution of Datasets

Runs	Training	Validation	Testing
Run1	1120 + OpenDeID v2 corpus	560 OpenDeID v2 corpus	950 OpenDeID v2 corpus
Run2	1120 + 614 OpenDeID v2 + 1304 i2b2 2014 + 1000 CEGS N-GRID corpus	560 OpenDeID v2 corpus	950 OpenDeID v2 corpus

2.2 Data Pre-processing

Before commencing the experiment, a critical initial step involves the pre-processing of the tab-separated annotation file to address the presence of the Byte Order Mark (BOM) character. This character is systematically removed through the utilization of Python, specifying the 'utf-8-sig' encoding to ensure the seamless reading of annotations. Subsequently, for compatibility with the deidentification pipeline, which requires input in XML format for training and validation, a conversion from TXT to XML is imperative. This conversion is executed via a Python script, utilizing a Document Type Definition (DTD) file and a mask annotation file to achieve a standardized and more comprehensible representation and organization of annotations. During the testing phase of the conversion code, we encountered encoding and annotation errors in certain files. These errors necessitated manual intervention to identify and rectify them, ensuring that the data integrity was maintained for further processing. This involved meticulous scrutiny and correction of errors to ensure the accuracy and reliability of the data for subsequent stages of analysis or use.

2.3 Deidentification Method

Extraction and deidentification of SHI were performed utilizing the OpenDeID pipeline [18]. This pipeline employs an innovative hybrid methodology that integrates associative rules, supervised deep learning, and pre-trained language models to achieve the deidentification of unstructured EHR text notes data from real-world Australian healthcare contexts. The algorithm's ingenuity is evident in its strategic utilization of diverse corpora, each with distinct settings, leading to the identification of the best-performing run characterized by superior accuracy. This approach represents a substantial leap forward in addressing the limitations of existing deidentification methodologies, illustrating the promising potential for real-time processing in the deidentification of EHR text notes within the healthcare domain. Apart from deidentification, the OpenDeID pipeline can generate surrogates [27]. Figure 1 represents an overview of theOpenDeID pipeline.

Segmentation, Tokenization, Tagging and Labeling
After completing the pre-processing procedures, the subsequent phase entailed the segmentation and tokenization of XML files, accomplished using the Spacy library. To ensure precise tokenization, systematic application of regular expressions was implemented. Following tokenization, the identified tokens were meticulously reintegrated into their corresponding sentences, forming a coherent representation of the processed

data. The deidentification process was intricately formulated as a sentence-tagging task, where the goal was to predict labels assigned to individual tokens within each sentence. These labels functioned as indicators for various categories of SHI. Embracing the BIESO tagging scheme, where "B," "I," and "E" signify the "beginning," "inside," and "ending" of a specific SHI, and "S" denotes a single-word SHI, the task aimed to discern and label the intricate nuances of SHI instances within the text. This methodology harnesses the structured labeling approach to precisely identify and categorize SHI entities within the sentences. The adoption of the BIESO tagging scheme ensures a detailed representation of SHI instances, capturing their contextual relationships and positional information within the text. This deliberate design not only enhances the interpretability of the deidentification process but also facilitates a more robust and nuanced prediction of SHI categories within the medical text data.

Experimental Setup
BERT, a specialized machine learning framework designed for applications in NLP, is grounded in the principles of deep learning models, particularly transformers. The transformer architecture consists of an encoder responsible for processing textual input and a decoder generating output based on predictions for the provided input. To tackle the deidentification task, an initially integrated pretrained BERT-base model was enhanced with an additional output layer for token-level classification. Fine-tuning of the model on labeled data, specifically discharge summaries in our experiments, was carried out using BioBERT initialization. The implementation of BERT-base models utilized PyTorch (version 1.10.1) in Python 3.9. A consistent batch size of 32 was employed throughout training, validation, and testing. Training concluded based on the cessation of improvement in either token-level accuracy or the micro-averaged F1-score in the validation set for a predetermined number of patience epochs. The training process spanned 20 epochs, with a patience setting of 5. The input sequence length for the BERT-base model was configured to 128, with sentences exceeding this length segmented into appropriate sections. In this experimental setup, emphasis was placed on discharge summary BioBERT, initialized from BioBERT and pretrained on discharge summaries. Notably, default hyperparameters were maintained without fine-tuning to ensure efficient training. The incorporation of BERT in this context underscores its adaptability to specialized tasks in NLP, with meticulous fine-tuning ensuring optimal performance in the deidentification task.

2.4   Evaluation Metrices

The evaluation of each SHI relies on three crucial attributes: the starting position, ending position, and category information of SHI within the text. These attributes serve as the foundation for defining True Positive (TP), False Positive (FP), and False Negative (FN) instances. 1. TP: Instances where the system correctly identifies SHI, meeting all three attributes (starting position, ending position, and category) as manually annotated in the dataset. 2. FP: Instances where the system incorrectly identifies attributes (start, end, or category) of SHI that do not align with any manually annotated results. 3. FN: Instances where manually annotated SHI cannot find a corresponding identification by the system.

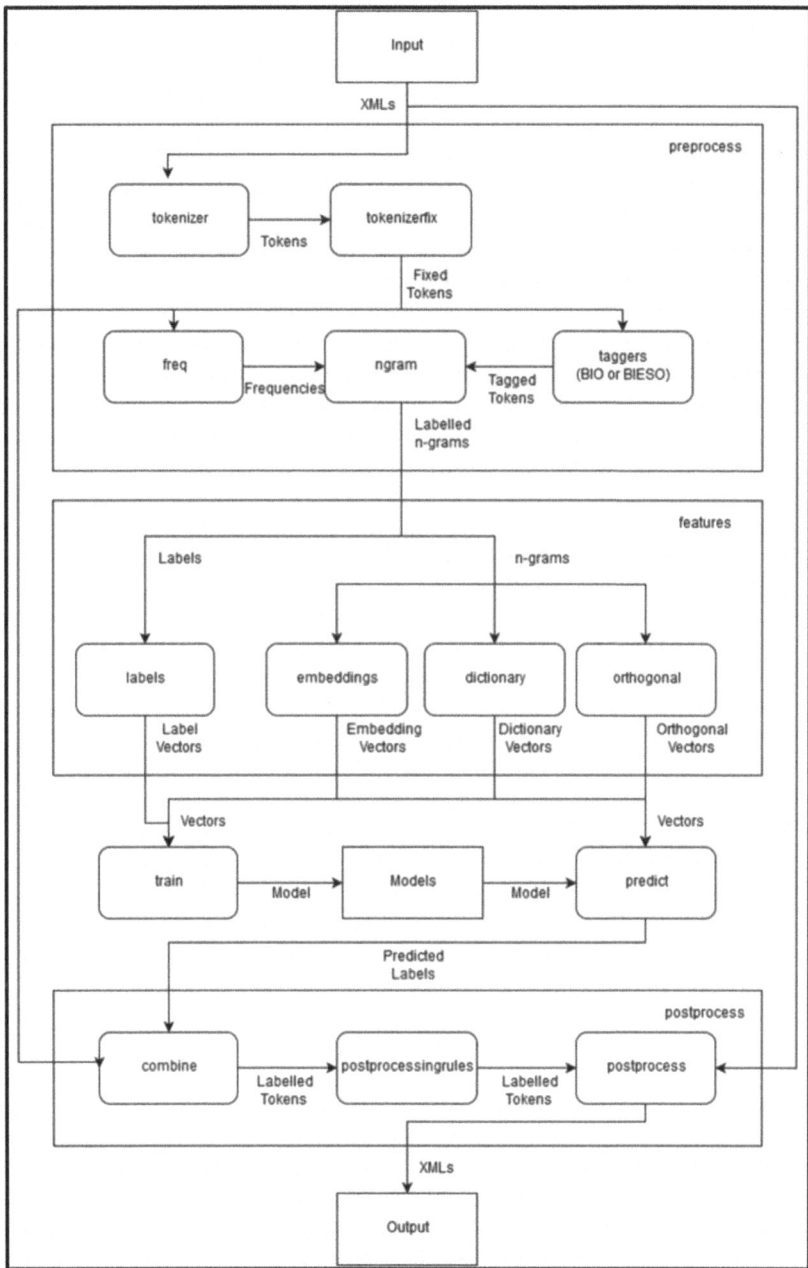

Fig. 1. Overview of OpenDeID pipeline

Using these definitions, the evaluation employs Precision, Recall, and F1-Measure metrics. Precision measures the proportion of correctly identified SHI instances out of all identified instances, while Recall calculates the proportion of correctly identified SHI instances out of all actual SHI instances. F1-Measure, which combines Precision and Recall into a single metric, provides a balanced assessment of system performance. Furthermore, Macro-F1-Measure serves as the ranking indicator for this subtask, considering the overall performance across all categories. This metric provides a comprehensive evaluation of the system's ability to correctly identify SHI instances across different categories, contributing to a more nuanced understanding of system effectiveness in preserving patient privacy within EHR text notes. Micro-averaged, macro-averaged precision, recall, and F1-scores are chosen for evaluating model performance in tasks like entity recognition in NLP because they provide a comprehensive view. Micro-averaging ensures fairness in imbalanced datasets, while macro-averaging reveals performance across all classes. Precision measures relevance, recall captures completeness, and F1-score balances both. Together, they offer a nuanced understanding crucial for informed decisions in model development and deployment. Micro-averaged precision, recall, and F1-scores aggregate the performance across all instances in the dataset, weighing each instance equally.

The Precision, Recall, Micro-F1, and Macro-F1 measures are calculated using the following formulas:

$$\text{Precision}_{(c)} = \frac{\text{True Positives}_{(c)}}{\left(\text{True Positives}_{(c)} + \text{False Positives}_{(c)}\right)} \tag{1}$$

$$\text{Recall}_{(c)} = \frac{\text{True Positives}_{(c)}}{\left(\text{True Positives}_{(c)} + \text{False Negatives}_{(c)}\right)} \tag{2}$$

$$\text{Micro-F}_1\text{-Measure}_{(c)} = 2 \cdot \frac{(P \cdot R)}{(P + R)} \tag{3}$$

$$\text{Macro F}_1\text{-Measure} = \frac{2 * \frac{\sum_{i=1}^{n} P_i}{|C|} * \frac{\sum_{i=1}^{n} R_i}{|C|}}{\frac{\sum_{i=1}^{n} P_i + \sum_{i=1}^{n} R_i}{n}} \tag{4}$$

Here, C represents the category, c represents the subcategory, and n represents the total number of occurring subcategories.

3 Results

Table 2 displays the performance metrics of the model across various configuration settings, including both Run1 and Run2. Run1 exhibits notably superior overall performance compared to other configurations. In this specific run, the model was trained on a dataset comprising 1120 reports, validated on 560 reports, and subsequently tested on 950 reports. The micro-averaged F1-scores for precision, recall, and F-measure were recorded as 0.913546, 0.8798339, and 0.8963734, respectively. Moreover, the macro-averaged F1-scores for precision, recall, and F-measure were calculated as 0.7596718, 0.6969127, and 0.7269402 respectively, providing additional insights into the model's performance across different categories. Comparatively, Run2 also produced good results, albeit slightly lower than Run1. In Run2, the model achieved micro-averaged

F1-scores of 0.9037531, 0.8330899, and 0.866984 for precision, recall, and F-measure, respectively. The macro-averaged F1-scores for precision, recall, and F-measure in Run2 were calculated as 0.6919585, 0.6262063, and 0.6574425, respectively. While Run2's performance is slightly inferior to Run1's, it still demonstrates the model's capability to effectively identify and remove sensitive information from EHR text notes. During the competition, our team, designated as TEAM_3902, secured an overall ranking of 52, with task-specific rankings of 38 and 97. This positioning reflects our team's performance relative to other participants and underscores our contribution to the competition's objectives of privacy protection and data standardization in medical records.

Table 2. Evaluation of OpenDeID pipeline in test sets

Runs		Strict Precision	Strict Recall	Strict F-measure
Run1	Micro-averaged	0.9135467	0.8798339	0.8963734
	Macro-averaged	0.7596718	0.6969127	0.7269402
Run2	Micro-averaged	0.9037531	0.8330899	0.866984
	Macro-averaged	0.6919585	0.6262063	0.6574425

4 Discussion

EHRs play a pivotal role in modern healthcare, offering a comprehensive repository of patient information that enables efficient management and dissemination of health data among healthcare practitioners. However, the widespread adoption of EHRs has raised concerns regarding patient privacy and the need to SHI contained within these records. The deidentification of EHR text notes [2] serves as a crucial process in preserving patient privacy while facilitating the responsible and ethical utilization of healthcare data for research, analysis, and clinical decision-making.

Preserving the confidentiality of SHI is paramount to uphold patient rights and maintain trust in the healthcare system. SHI encompasses a range of sensitive data, including patient names, addresses, dates of birth, medical history, and treatment details, the exposure of which could lead to breaches of privacy, identity theft, and other detrimental consequences for individuals. Therefore, effective deidentification techniques are essential to mitigate these risks [28] and ensure compliance with privacy regulations such as the Health Insurance Portability and Accountability Act (HIPAA) in the United States.

Our research endeavors to fulfill this pressing requirement by enhancing deidentification methods for electronic health record (EHR) text notes utilizing a hybrid approach [29]. By actively participating in both the AI Cup 2023 - Privacy Protection and Medical Data Standardization Competition and the 2024 International Workshop on Deidentification of Electronic Medical Record Notes (IW-DMRN), our objective was to enhance the efficacy of privacy preservation strategies in the utilization of healthcare data.

The results of our study demonstrate the efficacy of our approach in accurately identifying and removing SHI from EHR text notes while preserving data utility. Notably, our method leverages a hybrid deidentification pipeline that integrates associative rules, supervised deep learning, and pretrained language models [30–32]. This innovative approach allows for comprehensive processing of unstructured EHR text notes, leading to improved accuracy and efficiency in SHI extraction. Moreover, most existing methodologies involve training models on corpora derived from single healthcare centers or cohorts of patients with ailments. Consequently, there is limited empirical evidence regarding the practical application of these methods in real-time EHR text note deidentification processes. Previously, we employed the same hybrid deidentification pipeline, specifically tailored for the OpenDeID Corpus v1, which demonstrated promising results [18].

Furthermore, our method distinguishes itself from existing approaches through several key features. Firstly, the incorporation of diverse corpora, including the OpenDeID corpus v2, the 2014 i2b2/UTHealth deidentification corpus, and the 2016 CEGS N-GRID deidentification corpus, enhances the robustness and generalizability of our model. By training our model on datasets sourced from multiple healthcare institutions [33], we ensure its effectiveness across varied contexts and data types. Secondly, our method utilizes a structured BIESO tagging scheme for SHI identification, enabling precise labeling and categorization of SHI entities within EHR text notes. This structured approach enhances the interpretability and reliability of our model's predictions, ensuring accurate identification of SHI categories and their contextual relationships within the text. By leveraging pre-trained language models and specialized architectures, we achieve state-of-the-art results in SHI extraction, demonstrating the potential for real-time processing in healthcare settings.

While the OpenDeID pipeline demonstrates considerable promise in addressing the challenges of deidentification, it is not without its limitations. One notable constraint is its reliance on predefined rules and patterns, which may not always capture the nuanced complexities present in diverse healthcare datasets. Additionally, the performance of the pipeline can be influenced by variations in data quality, such as inconsistencies in formatting and language use across different EHR systems. Furthermore, the scalability of the pipeline may be hindered when dealing with large-scale datasets or when confronted with evolving privacy regulations and standards. As such, continuous refinement, and adaptation of the OpenDeID pipeline are necessary to overcome these limitations and ensure its effectiveness in real-world healthcare settings.

5   Conclusion

Our study on deidentifying SHI across multicenter corpora using a hybrid deidentification pipeline highlights the paramount significance of safeguarding patient privacy while harnessing EHR text notes. This approach not only ensures the protection of sensitive patient data but also facilitates the utilization of EHR information to improve healthcare outcomes. By employing advanced deidentification techniques across diverse healthcare settings, our study aims to enhance patient privacy and contribute to the realization of improved healthcare outcomes, such as enhanced clinical decision-making, personalized treatment plans, and more efficient healthcare delivery..Our participation in the AI Cup 2023 - Privacy Protection and Medical Data Standardization Competition and the 2024 International Workshop on Deidentification of Electronic Medical Record Notes

(IW-DMRN) reflects our commitment to refining techniques for safeguarding privacy in healthcare data utilization. By employing a hybrid approach to deidentification across diverse datasets has yielded significant contributions to the advancement of comprehension and strategies for safeguarding sensitive information within varied datasets. This innovative methodology has provided invaluable insights into the intricate nature of protecting confidential data across different data sources and healthcare settings. By combining rule-based systems with state-of-the-art machine learning techniques, our study has shed light on the complex interplay between data characteristics and deidentification strategies. Furthermore, our findings offer novel perspectives on the optimal integration of multiple approaches to enhance the robustness and effectiveness of privacy protection measures. Through this hybrid approach, we have not only deepened our understanding of the challenges associated with safeguarding sensitive information but also paved the way for the development of more sophisticated and adaptable deidentification frameworks tailored to diverse datasets and healthcare contexts. Utilizing the OpenDeID pipeline and incorporating the BERT-base model fine-tuned on discharge summaries, our study has demonstrated superior performance in accurately deidentifying electronic health record text notes. The achieved precision, recall, and F-measure scores underscore the efficiency and reliability of our model in precise deidentification tasks. Overall, our research highlights the ethical and privacy-conscious utilization of healthcare data within dynamic multicenter healthcare frameworks. Moving forward, our findings serve as a cornerstone for further advancements in deidentification techniques, promoting the responsible and ethical use of healthcare data for improved patient care and healthcare outcomes.

Acknowledgments. We would like to express our heartfelt gratitude to the organizers of the AI CUP 2023 Autumn Competition (https://codalab.lisn.upsaclay.fr/competitions/15425), overseen by the Artificial Intelligence Competition and Annotation Data Collection Project Office of the Ministry of Education in Taiwan. Our appreciation also goes to the University of New South Wales, Lowy Cancer Research Centre, and Health Science Alliance Biobank for granting access to the medical records utilized in this study. Additionally, we recognize the IW-DMRN workshop (https://www.sredhconsortium.org/sredh-competitions/sredhai-cup-2023/2024-iw-dmrn) for their valuable contributions to our research endeavors and the SREDH Consortium (https://www.sredhconsortium.org/) Translational Cancer Bioinformatics working group in accessing the OpenDeID corpus v2 Dataset. JJ was funded by the Australian National Health and Medical Research Council (GNT1192469) and supported by 2022 Google's research grants and cloud computing resources (GCP19980904), as well as the Research Technology Services at University of New South Wales Sydney and NVIDIA's academic hardware grant programs.

Disclosure of Interests. The authors have no competing interests to declare that are relevant to the content of this article.

References

1. Forde-Johnston, C., Butcher, D., Aveyard, H.: An integrative review exploring the impact of Electronic Health Records (EHR) on the quality of nurse-patient interactions and communication. J. Adv. Nurs. **79**(1), 48–67 (2023)

2. Yoon, J., et al.: EHR-safe: generating high-fidelity and privacy-preserving synthetic electronic health records. NPJ Digit. Med. **6**(1), 141 (2023)
3. Lombardo, G., et al.: Electronic health records (EHRs) in clinical research and platform trials: application of the innovative EHR-based methods developed by EU-PEARL. J. Biomed. Inform. **148**, 104553 (2023)
4. Adamson, B., et al.: Approach to machine learning for extraction of real-world data variables from electronic health records. Front. Pharmacol. **14**, 1180962 (2023)
5. Clusmann, J., et al.: The future landscape of large language models in medicine. Commun. Med. **3**(1), 141 (2023)
6. Juhn, Y., Liu, H.: Artificial intelligence approaches using natural language processing to advance EHR-based clinical research. J. Allergy Clin. Immunol. **145**(2), 463–469 (2020)
7. Labinsky, H., et al.: An AI-powered clinical decision support system to predict flares in rheumatoid arthritis: a pilot study. Diagnostics **13**(1), 148 (2023)
8. Hossain, E., et al.: Natural language processing in electronic health records in relation to healthcare decision-making: a systematic review. Comput. Biol. Med. **155**, 106649 (2023)
9. Catelli, R., Esposito, M.: Chapter 7 - De-identification techniques to preserve privacy in medical records. In: Chatterjee, P., Esposito, M. (eds.) Artificial Intelligence in Healthcare and COVID-19, pp. 125–148. Academic Press (2023)
10. El-Hayek, C., et al.: An evaluation of existing text de-identification tools for use with patient progress notes from Australian general practice. Int. J. Med. Inform. **173**, 105021 (2023)
11. Radhakrishnan, L., et al.: A certified de-identification system for all clinical text documents for information extraction at scale. JAMIA Open **6**(3), ooad045 (2023)
12. Xiao, Y., et al.: In the name of fairness: assessing the bias in clinical record de-identification. In: Proceedings of the 2023 ACM Conference on Fairness, Accountability, and Transparency, pp. 123–137. Association for Computing Machinery, Chicago, IL, USA (2023)
13. Lee, H.J., et al.: A hybrid approach to automatic de-identification of psychiatric notes. J. Biomed. Inform. **75s**, S19–S27 (2017)
14. Zhao, Y.S., et al.: Leveraging text skeleton for de-identification of electronic medical records. BMC Med. Inform. Decis. Mak. **18**(Suppl 1), 18 (2018)
15. ISLAB. Privacy protection and medical data standardization competition: decoding clinical cases and letting data tell stories (2023). https://codalab.lisn.upsaclay.fr/competitions/15425
16. Mir, T.H., et al.: Deidentification and temporal normalisation of the electronic health record notes using large language models: the SREDH/AI-Cup 2023 deidentification competition. In: 2024 International Workshop on Deidentification of Electronic Medical Record Notes. Springer, Kaohsiung (2024)
17. SREDH Consortium: 2024 International workshop on deidentification of electronic medical record notes (IW-DMRN) (2024). https://www.sredhconsortium.org/sredh-competitions/sredhai-cup-2023/2024-iw-dmrn
18. Liu, J., et al.: OpenDeID pipeline for unstructured electronic health record text notes based on rules and transformers: deidentification algorithm development and validation study. J. Med. Internet Res. **25**, e48145 (2023)
19. Gupta, S., et al.: Preliminary evaluation of fine-tuning the OpenDeID deidentification pipeline across multi-center corpora. In: 34th Medical Informatics Europe Conference (MIE2024), Athens, Greece (2024)
20. Alla, N.L.V., et al.: Cohort selection for construction of a clinical natural language processing corpus. Comput. Methods Programs Biomed. Update **1**, 100024 (2021)
21. Jonnagaddala, J., et al.: The OpenDeID corpus for patient de-identification. Sci. Rep. **11**(1), 19973 (2021)
22. SREDH Consortium: Secure Research Environment for Digital Health (SREDH) Consortium. https://www.sredhconsortium.org/

23. Stubbs, A., Kotfila, C., Uzuner, Ö.: Automated systems for the de-identification of longitudinal clinical narratives: overview of 2014 i2b2/UTHealth shared task Track 1. J. Biomed. Inform. **58**(Suppl), S11–S19 (2015)
24. Stubbs, A., Uzuner, Ö.: Annotating longitudinal clinical narratives for de-identification: the 2014 i2b2/UTHealth corpus. J. Biomed. Inform. **58**(Suppl), S20–S29 (2015)
25. Stubbs, A., Filannino, M., Uzuner, Ö.: De-identification of psychiatric intake records: overview of 2016 CEGS N-GRID shared tasks Track 1. J. Biomed. Inform. **75s**, S4–S18 (2017)
26. i2b2. NLP Research Data Sets. https://www.i2b2.org/NLP/DataSets/
27. Chen, A., et al.: Generation of surrogates for de-identification of electronic health records. Stud. Health Technol. Inform. **264**, 70–73 (2019)
28. Fernandes, A.C., et al.: Development and evaluation of a de-identification procedure for a case register sourced from mental health electronic records. BMC Med. Inform. Decis. Mak. **13**(1), 71 (2013)
29. Ahmed, T., Aziz, M.M.A., Mohammed, N.: De-identification of electronic health record using neural network. Sci. Rep. **10**(1), 18600 (2020)
30. Abdalla, M., et al.: Exploring the privacy-preserving properties of word embeddings: algorithmic validation study. J. Med. Internet Res. **22**(7), e18055 (2020)
31. Devlin, J., et al.: BERT: pre-training of deep bidirectional transformers for language understanding. In: North American Chapter of the Association for Computational Linguistics (2019)
32. Jin, M., et al.: A hybrid machine learning method for the de-identification of un-structured narrative clinical text in multi-center Chinese electronic medical records data. In: 2019 IEEE International Conference on Big Knowledge (ICBK) (2019)
33. Kushida, C.A., et al.: Strategies for de-identification and anonymization of electronic health record data for use in multicenter research studies. Med. Care **50**(Suppl), S82–S101 (2012)

Sensitive Health Information Extraction from EMR Text Notes: A Rule-Based NER Approach Using Linguistic Contextual Analysis

Ming-Sheng Huang , Bo-Ren Mau$^{(\boxtimes)}$, Jie-Hui Lin , and Ying-Zhen Chen

National Kaohsiung University of Science and Technology, No.1, University Road, Yanchao, Kaohsiung 82445, Taiwan
{C110133204,brnmau,C110118110,C110118213}@nkust.edu.tw

Abstract. The rapid digitization of electronic medical record (EMR) text notes has notably improved data analysis and patient care but also introduced data privacy challenges. Named Entity Recognition (NER) plays a vital role in protecting sensitive health information (SHI) during healthcare data analytics. This paper introduces a rule-based NER system designed to identify SHI entities within EMR text notes. Compared to machine learning-based (ML) systems, our approach efficiently adapts to the dynamic medical language and new terminologies in EMR text notes. The system primarily focuses on extracting SHI entities and normalizing temporal entities. We developed a rule-based NER system using a dataset of 1,734 training and 560 testing EMR text notes. The system employs HashMap and HashSet to categorize 18 SHI entities. The system features an information extraction algorithm that integrates contextual and linguistic rules analysis for precise and adaptable recognition. Our approach includes iterative optimization for refining the entity-specific patterns and improving contextual rules. Performance is evaluated through precision, recall, and *F1*-score metrics. Our results demonstrate an iterative optimization process of updating linguistic rules contributes to the system's enhanced performance, increasing the *F1 score* from 0.63 to 0.92. In SHI entity recognition, the system achieved an average *F1 score* of 0.93. For temporal entity normalization, the system maintained high performance with an average *F1*-score of 0.90. In an ML-dominated NER field, our rule-based system not only demonstrates the adaptability and effectiveness of rule-based approaches but also provides a framework that effectively integrates new linguistic rules. Additionally, it highlights the potential of hybrid models, combining rule-based and ML methods for future research. This could lead to improved pattern recognition and increased efficiency in EMR text notes.

Keywords: Named Entity Recognition · Temporal Entity Normalization · Rule-Based Systems · Linguistic Pattern Recognition · Information Extraction

J. Jonnagaddala et al. (Eds.): IW-DMRN 2024, CCIS 2148, pp. 120–133, 2025.
https://doi.org/10.1007/978-981-97-7966-6_9

1   Introduction

1.1 Background

As healthcare facilities transition from traditional paper-based records to electronic medical records (EMR) text notes, these digital records encompass a wealth of medical information, including patient symptoms, diagnoses, and medical observations [1]. The effective utilization of data from EMR text notes has the potential to assist in disease registries, epidemiological studies, drug safety surveillance, clinical trials, and health audits [2, 3].

1.2 Named Entity Recognition in EMR Text Notes

NER in the medical domain serves as a fundamental task for extracting meaningful entities, especially sensitive health information (SHI), from EMR text notes [4]. This extraction is crucial for informing critical downstream tasks such as clinical assertion status and the de-identification of sensitive data [5, 6]. The literature reveals two primary NER approaches: rule-based and ML-based approaches. These approaches have distinct methodologies and applications, offering different advantages and challenges in the context of EMR text notes data.

1.3 Rule-Based Approach in NER

Information extraction (IE) is the process of automatically identifying and extracting specific pieces of information, such as named entities [7]. In recognizing and extracting the named entities, the rule-based approach plays a crucial role in named entity IE [8–11]. Rule-based NER has been widely implemented using regular expressions (RE) or dictionary information, incorporating characteristics of the entities [9]. As Robredo (1982) asserts, in any area of knowledge, meaningful terms can be used as descriptors to represent the content of written documents for constructing rules. This approach has been exemplified in various applications, such as the rule-based named entity recognizer, the suffix-based text segment matching algorithm for NER, and the dependency path-based relationship extraction matching algorithm, for semantic web IE [10, 11].

For example, Spositto et al. (2021) [7] proposed a rule-based approach for IE from legal texts using RE to identify named entities. The defined patterns using RE to recognize entities such as File No., Resolution No., and Article No. of Law XXX, common in legal documents. The experimental results demonstrated the effectiveness of the rule-based NER approach for IE in the legal domain. While the rule-based approach in NER demonstrates advantages and exceptional performance, its limitation lies in the lack of adaptability of rules to new entities and lexicons. This limitation constrains the approach's generalization to other domains [12].

However, Oudah and Shaalan (2016) [4] used part-of-speech tags to formulate new linguistic rules, aiming to improve the recognition of named entities and thereby enhance the performance of the rule-based component in NER systems under the constraint. Similarly, Eftimov et al. (2017) [9] suggested the use of syntactic analysis to improve performance within this limitation by understanding grammatical structures and identifying relationships between different words and phrases.

In the context of EMR text notes, this approach can adapt predefined rules and terminological resources. It performs well where language is specialized and well-defined, allowing for precise targeting of specific linguistic patterns and medical terminologies. These can be frequently updated to reflect new medical knowledge without requiring extensive retraining [4].

1.4   Machine Learning-Based Approach in NER

ML-based NER involves using machine learning techniques to identify and classify named entities in an annotated corpus [13, 14]. Commonly employed ML algorithms for NER include Hidden Markov Models (HMM) [15], Long Short-Term Memory (LSTM) [16], and Conditional Random Fields (CRF) [17]. For example, Cao (2019) applied the CRF algorithm to train on 1000 EMR text notes, focusing on hospital name entities [18]. Chen (2017) developed an LSTM model to create word embeddings and character representations for the extraction of entities [19].

In comparison to rule-based approaches, ML-based approaches are less reliant on terminological resources and manually created rules. However, they have limitations in adapting to the rapidly changing medical language [4]. This approach generally requires extensive, up-to-date annotated corpora. The requirement to retrain models for new terms or linguistic patterns can be resource-intensive, demanding substantial effort from human experts. Moreover, it is conceivable that the rule-based approach could outperform ML-based approaches in NER for EMR text notes. This could be attributed to the domain knowledge and expertise of human experts in understanding the context and nuances of EMR text notes, which enables them to accurately identify, and label named entities. Human experts also can iteratively refine and improve the rule-based approaches based on their experience and feedback, which may lead to enhanced performance [20].

1.5   The Dynamic Nature of Language in EMR Text Notes

The language in EMR text notes is characterized by its dynamic nature, continuously evolving with new medical discoveries and terminology [4]. This dynamism, primarily manifested in evolving terminology rather than structural changes, makes rule-based approaches particularly advantageous. Rule-based approaches can be rapidly updated to reflect new medical terms and concepts. In contrast, ML methods, despite their ability to handle large datasets, often struggle with the agility required to adapt to these frequent and specific terminological updates [9].

1.6   Research Purpose and Research Questions

The limitations associated with the rule-based approach, as identified in the previous studies, have the potential for enhancement. Therefore, in this study, we propose a different method that more closely aligns with the nature of language: contextual analysis. This innovative rule-based NER system, designed specifically for use with EMR text notes, aims to develop a more effective and efficient method for extracting SHI entities across various domains. We deployed this system in the AI-Cup 2023 competition [21,

30], achieving an overall rank of 1, and task-wise ranks of 2 for Task 1 and 1 for Task 2, under the team's name C110133204 with the team code Team_4761. The dataset used in this study, OpenDeID, was surrogated by Chen et al. (2019) [27]. It was specifically employed in the AI-Cup 2023 competition and IW-DMRN workshop [31] of SREDH [32]. We aim to address two main research questions:

1. How effective is the performance of our rule-based NER system for SHI entities recognition?
2. How effective is the normalization of temporal-related SHI entities?

2   Method

2.1   OpenDeID v2 Corpus

The data for this study originates from the pathology reports at the Health Science Alliance Biobank, Lowy Cancer Research Centre, University of New South Wales, Australia [22]. OpenDeID v2 corpus consists of a total of 3,244 pathology reports from which we used a dataset comprising 1,734 training EMR text notes with annotation files and 560 testing EMR text notes to construct and evaluate our system. The dataset includes 18 SHI entities, such as AGE, DOCTOR, CITY, and COUNTRY. Alla et al. (2021) [28] have identified several challenges associated with this corpus. For example, the format of EMR text notes varies across different hospitals, presenting challenges for data processing.

2.2   Entity Categorization Using HashMap and HashSet

To address the challenges posed by the diversity in EMR text note formats, this section outlines a two-step process for categorizing annotation files to generate a collection file, which includes the use of HashMap and HashSet. This process aims to enhance data organization and retrieval efficiency, ensuring accurate and streamlined analysis of entities within the annotation files (Fig. 1).

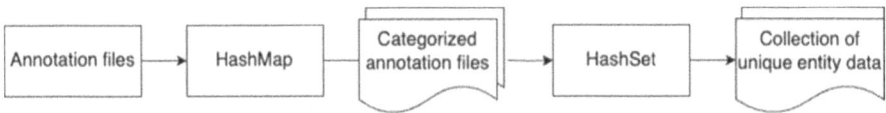

Fig. 1. Two-step process for entity categorization using HashMap and HashSet.

Entity Categorization with HashMap. A HashMap is a common data structure used to store key-value pairs for efficient data retrieval [23, 24]. Initially, we read all entities from the annotation file, designating them as keys. Subsequently, we created a HashMap to systematically categorize each entity. This process resulted in a series of categorized annotation files. Each file contains all entity elements related to a specific entity type, thereby facilitating organized data retrieval.

Unique Entity Identification with HashSet. Utilizing HashSet, an unordered collection of unique elements, we stored unique entity data extracted from the categorized annotation files. This step ensures that unique elements are retained, effectively eliminating repetitions. One of the key benefits of using HashSet in this context is its time efficiency. Operations such as adding, checking for presence, and removing elements can typically be performed in constant time, with a time complexity is O(1) [25]. This level of efficiency is crucial for processing large datasets, as it greatly reduces the computational time required for identifying and ensuring the uniqueness of entities in our data.

This step aims to employ HashSet to determine the degree of variation in the entity types. To calculate this variation, we use the formula $VR = \frac{N}{U} \times 100\%$, where N is the number of entity elements in the categorized file, and U is the number of unique entity elements after applying the HashSet process. The variation threshold is set at the 25% top quartile. A variation rate exceeding this threshold indicates a high variation rate in the entity elements, suggesting less fixed rules for that entity type.

For instance, in the categorized annotation file for the "STATE" entity, there are 428 elements. After applying the HashSet process, 16 unique elements emerge, indicating a variation rate of 3.7% for the "STATE" entity. This rate is below our cut-off threshold, suggesting that the form of the "STATE" entity is relatively fixed.

2.3   Development of the Rule-Based Information Extraction Algorithm

Figure 2 shows the algorithm for our rule-based NER system, detailed in Sect. 2.4. The algorithm includes two steps, which are analyzing contextual rules and constructing linguistic rules.

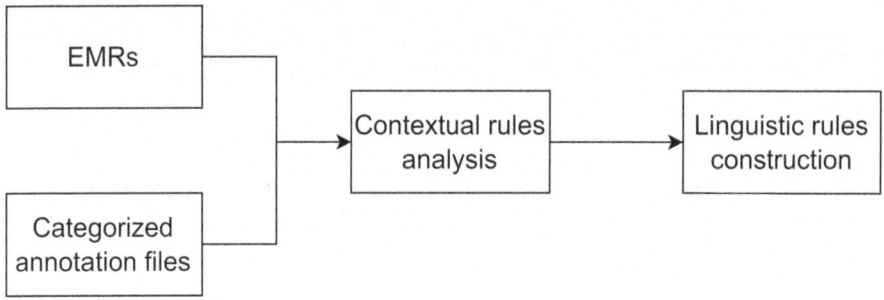

Fig. 2. Process of the rule-based IE algorithm.

Contextual Rules Analysis. Initially, we accessed the EMR text notes data and read the categorized annotation file. We match the file numbers, starting, and ending positions of each entity with corresponding information in the EMR text notes. Afterward, we extract the relevant information within the area and its contextual information to generate data contextually relevant for further analysis. For instance, the entry "102 DOCTOR 1433 1443 R Imaizumi" indicates that the doctor's name "R Imaizumi" appears in the EMR text notes of file number 102, and the starting and ending position indices are 1433 and 1443. This allows us to precisely locate the data and its context, such as "Result to Dr KURTIS HALDERMAN by Dr R Imaizumi at 12:00pm 14.5.12". With this data, we can more accurately analyze the rules governing this entity.

Linguistic Rules Construction. Secondly, we search for specific keywords within EMR text notes to compare them with the data identified through contextual rules analysis. This comparison aims to find connections. Next, we apply RE to search the EMR text notes for data matching specific patterns. This helps in determining if the specific pattern is associated with the entity. For entities with specific formats, we encapsulate their contextual linguistic rules into a RE rule database.

For instance, identifying the phrase "69-year-old female with adnexal mass for TAH" suggests "year old" as a keyword for the AGE entity. To confirm this, we identify all data in the EMR text notes containing this keyword and compare it with the data from contextual rules analysis. The prefix "year old" is likely relevant to the AGE entity. Based on this observation, we construct a RE rule as "(\d+) \syear old."

2.4   Architecture of the Rule-Based NER System

This section presents the architecture of our rule-based NER system, as shown in Fig. 3. It outlines the comprehensive steps involved, including the IE algorithm, time normalization, and the system's evaluation process.

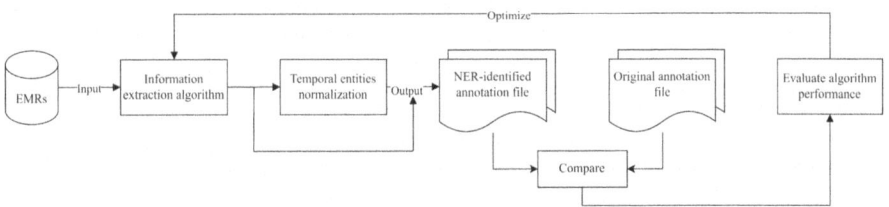

Fig. 3. Architecture of the rule-based NER system.

Information Extraction Algorithm. The IE process begins with the sequential reading of EMR text notes data as input. Upon loading the data, the IE algorithm is activated to invoke the RE rule database. This database contains an extensive collection of patterns that the algorithm employs to identify and classify various medical entities within the EMR text notes. By matching these patterns against the text in the EMR text notes, the algorithm efficiently locates and extracts the relevant entities. Once an entity is recognized, the algorithm cross-references it with the contextual rules established in the database, ensuring each entity's accuracy and relevance within the medical context of the EMR text notes.

Temporal Entities Normalization Process. For time-related entities, accurate identi-fication and standard format representation are critical. Upon discovering information within the time entity, the system first deconstructs it to understand its intended mean-ing. Following this, the entity is reformatted into a pre-determined format. For instance, the date '62.5.18' is parsed into (62)(5)(18), interpreted as May 18, 2062. It is then standardized to the format '2062-05-18' to complete temporal normalization.

System Evaluation and Optimization Process. In this step, we verify the accuracy of entity recognition by comparing the NER-identified annotation file against the original annotation file. It sequentially compares the entities of both files to pinpoint discrep-ancies. To evaluate the performance of our NER system, we adopted precision, recall, and $F1$-score as our evaluation metrics, following Powers (2020) [26]. Through itera-tive refinement based on contextual analysis, we progressively optimize the system's accuracy, aiming for the highest achievable accuracy rate.

Demonstration of Entity Extraction. An example of RE rule database for extracting CITY, STATE, and ZIP is shown in Algorithm 1 below. An example from the EMR text notes is: "ALDGAT Northern Territory 2263." This string encapsulates the entities CITY, STATE, and ZIP. Our analysis interprets STATE as state names and abbreviations, CITY as a preceding single uppercase English word, and ZIP as a post-state four-digit number.

In this text's processing, the term "Northern Territory" is recognized as the STATE entity, serving as a boundary for the CITY and ZIP entities. The IE algorithm applies a rule whereby the beginning of "Northern Territory" signifies the end boundary of CITY, and its end marks the start boundary of ZIP. A search operation identifies "ALDGAT" and "2263" as the respective prefix and suffix to STATE. RE matching verifies "ALDGAT" as an uppercase English word (CITY) and "2263" as a four-digit integer (ZIP). By applying the RE rule to identify STATE and using it as an anchor, the algorithm delineates the boundaries for CITY and ZIP extraction. The final validation through the RE rule ensures the correct format, resulting in the accurate identification of entities: CITY as "ALDGAT," STATE as "Northern Territory," and ZIP as "2263" (Fig. 4).

Algorithm 1 Entity Extraction from an EMR

1: Data: text_string containing CITY, STATE, and ZIP
2: Result: Extracted entities CITY, STATE, and ZIP
3: **procedure** EXTRACTENTITIES(text_string)
4:     Define state_regex as a pattern for state names and abbreviations
5:     Define city_regex as a pattern for a single uppercase word
6:     Define zip_regex as a pattern for a four-digit number
7:
8:     Set state_match to the result of searching state_regex in text_string
9:     **if** state_match is found **then**
10:         Set city_boundary_end to the start position of state_match
11:         Set zip_boundary_start to the end position of state_match
12:
13:         Set city_match to the result of searching city_regex before city_boundary_end
14:         Set zip_match to the result of searching zip_regex after zip_boundary_start
15:         **if** city_match and zip_match are found **then**
16:             Validate city_match and zip_match
17:             **if** city_match and zip_match are valid **then**
18:                 Set CITY to city_match
19:                 Set STATE to state_match
20:                 Set ZIP to zip_match
21:             **end if**
22:         **end if**
23:     **end if**
24:
25:     return CITY, STATE, ZIP
26: **end procedure**

Fig. 4. Example of entity extraction algorithms applied to an EMR text note.

3   Result

3.1   Effectiveness of SHI Entities Recognition in Iterative Optimization

As shown in Table 1, our approach achieved a precision of 0.95, recall of 0.90, and an $F1$-score of 0.93 for SHI entities recognition. Entities such as COUNTRY, HOSPITAL, and PHONE obtained the highest scores due to their clear rules and patterns. Conversely, the entities with the lowest scores were ORGANIZATION and PATIENT, as they lacked distinct rules and patterns for identification.

Table 1. Precision, Recall and *F1*-score metrics for SHI entity recognition.

Entity	Precision	Recall	*F1*-score
AGE	1.00	0.91	0.95
CITY	1.00	0.93	0.96
COUNTRY	1.00	1.00	1.00
DATE	0.98	0.82	0.89
DEPARTMENT	0.80	0.99	0.89
DOCTOR	0.99	0.92	0.96
DURATION	0.83	0.83	0.83
HOSPITAL	1.00	1.00	1.00
IDNUM	0.99	0.97	0.99
LOCATION-OTHER	1.00	0.75	0.86
MEDICALRECORD	0.99	0.94	0.97
ORGANIZATION	0.67	0.50	0.57
PATIENT	0.96	0.96	0.57
PHONE	1.00	1.00	1.00
STATE	0.99	0.99	0.99
STREET	0.97	0.99	0.98
TIME	0.95	0.78	0.86
ZIP	0.99	0.98	0.99
Average	0.95	0.90	0.93

Moreover, Fig. 5 shows the progression of our optimization performance across each iterative process. Initially, the first iteration yielded a precision of 0.87, recall of 0.49, and an *F1*-score of 0.63. The second iteration showed an improvement with a precision of 0.75, recall of 0.83, and an *F1*-score of 0.79. The third iteration further improved to a precision of 0.90, recall of 0.85 and an *F1*-score of 0.88. The final iteration culminated in a precision of 0.95, recall of 0.88 and an *F1*-score of 0.91.

In the initial optimization phase, we observed prevalent data formats within various entities, such as repetitive patterns for COUNTRY, STATE, and DEPARTMENT, along with fixed patterns for IDNUM and PHONE. Specifically, for the AGE entity, variations such as 'yr old' and 'year old' were identified as patterns. The precision for most entities was commendable, exceeding 0.9, while the recall spanned a wider range, from 0.4 to 0.7, leading to an average *F1*-score of 0.63. This result indicates that the entity recognition rules derived from these patterns are mostly accurate, yet there are still rules based on undiscovered patterns that need to be identified. For most entities, additional analysis and the incorporation of unprocessed rules based on undetected patterns are required for optimization.

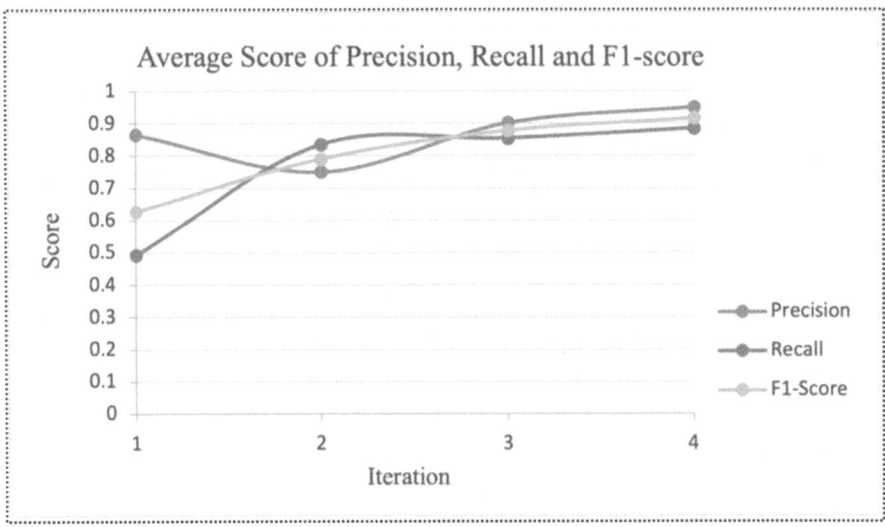

Fig. 5. Performance across iterative optimization phases of NER system.

Further iteration incorporated additional entity-specific patterns, which improved the rule set for each entity. For instance, patterns such as 'yo,' 'F,' and 'y.o' were added to the rule set for the AGE entity. The RE rule database for each entity was also augmented with external data that closely aligns with the identified patterns, such as adding a broader range of country names to the rule set for the COUNTRY entity. Following the second optimization, a notable increase in recall was observed across entities, with values between 0.7 and 0.99. However, this was countered by a reduction in precision for some entities, signaling an influx of false positives due to overgeneralized rules. Despite this, the average $F1$-score improved to 0.79, reflecting a more accurate identification of true positives as the rules were refined to align more closely with the specific patterns of each entity.

Continued optimization entailed targeted adjustments to the rule set to minimize false positives. For instance, in the case of the AGE entity, the suffix 'F' was initially matched by a rule that looked for one to two preceding digits. However, this rule was too broad and resulted in numerous false positives. To rectify this, we refined the rules by adding conditions that excluded alphabetical characters in the suffix 'F' and limited the preceding integers to a maximum of two digits. These rule refinements greatly enhanced accuracy, culminating in an improved average $F1$-score of 0.88 during the third optimization phase.

Despite these improvements, precise recognition for certain entities, such as ORGA-NIZATION, remained elusive due to the lack of specific identifiable patterns and corresponding rules for accurate processing. Additionally, some entities, even with well-defined rules, continued to generate a number of false positives. To address this, we iteratively modified the existing rules, methodically evaluating their impact on both false positives and negatives with each iteration. This optimization process eventually increased the overall average $F1$-score to 0.91.

3.2 Effectiveness of Temporal-Related SHI Entities Normalization

The DATE entity's $F1$-score in SHI recognition achieved 0.89, while in temporal normalization, it decreased to 0.82. This discrepancy suggests that while our system is adept at identifying temporal SHI entities, there are shortcomings in the normalization process. The main issue leading to this error is the misidentification of some DATE entity data, such as the year '2062', which could be confused with a ZIP entity. This ambiguity has led to partial inaccuracies in the normalization process (Table 2).

Table 2. Precision, Recall, and $F1$-score for temporal normalization.

Entity	Precision	Recall	$F1$-score
DATE	0.84	0.81	0.82
TIME	0.95	0.75	0.84
DURATION	1.00	0.91	0.95
Average	0.94	0.86	0.90

4 Conclusion

In the realm of NER, where ML approaches are prevalent, our rule-based NER system underscores the importance of adaptability to the dynamic nature of medical language. This system has demonstrated not only the feasibility of rule-based methods in an ML-dominated NER field but also the potential for dynamically updating linguistic patterns. With an average $F1$-score of 0.91 across our two main tasks, the results affirm the effectiveness of rule-based strategies in capturing the evolving medical lexicon.

5 Discussion

Liu et al. (2023) [29] combined rule-based and ML-based approaches to achieve the $F1$-score of 0.9659 for extracting SHI in EMR text notes. They used the SpaCy toolkit to segment and assign tokens in the SHI. However, some tokens were incorrectly cascaded, which can have an impact on the SHI tagging. To solve this problem, they employed RE. Compared to their method, our method eschews the SpaCy toolkit for segmentation to streamline our process with a focus on RE from the outset. By directly applying RE, we efficiently refine data extraction and minimize the occurrence of incorrectly cascaded tokens, achieving an $F1$-score of 0.91 without using an ML-based approach.

Our current IE system employs a brute-force search approach, which is extremely time-consuming. It sequentially searches from the beginning to the end of the text note, becoming particularly inefficient when dealing with lengthy EMR text notes. To address this issue, we propose the future adoption of a two-pointer algorithm. This algorithm promises to enhance efficiency by simultaneously searching for entities using right and left pointers within an array, eliminating the dependency on sequential searches. By implementing this strategy, our IE system is expected to deliver considerably faster performance. Additionally, we intend to integrate the methods advocated by Oudah and Shaalan (2016) [4] and Eftimov et al. (2017) [9] into our NER system. This integration involves the utilization of part-of-speech tags and syntactic analysis, aiming to facilitate substantial contributions across diverse domains.

Moreover, we suggest future research to explore hybrid systems that combine rule-based and ML approaches. Such systems could further contribute to enhancing the automatic identification of novel linguistic patterns, thereby improving the efficiency and accuracy of NER applications in healthcare. Furthermore, we propose conducting comprehensive evaluations of our system, with an emphasis on contextual analysis, across different domains within NER to achieve more precise results in future studies.

Acknowledgments. We extend our sincere thanks to the organizer of the AI CUP 2023 Autumn Competition, managed by the Artificial Intelligence Competition and Annotation Data Collection Project Office of the Ministry of Education, Taiwan. We are also grateful to the University of New South Wales, Lowy Cancer Research Centre, and Health Science Alliance Biobank for providing the medical records analyzed in this study. Furthermore, we acknowledge the IW-DMRN workshop and the SREDH Consortium for their contributions to our research efforts.

Disclosure of Interests.   The authors hereby declare that there are no competing interests, financial or otherwise, that could have influenced the work reported in this research. This assertion is made with the understanding that competing interests, as defined, pertain to relationships or actions that could potentially bias or be perceived to bias the work's objectivity, integrity, or interpretation. Thus, it is affirmed that the content of this article is presented in an unbiased manner and is the result of independent work, free from any external influence that could compromise the scientific integrity of the research presented.

References

1. Ngiam, K.Y., Khor, I.W.: Big data and machine learning algorithms for health-care delivery. Lancet Oncol. **20**(5), e262-273 (2019). https://doi.org/10.1016/s1470-2045(19)30149-4
2. Ford, E., Carroll, J.A., Smith, H.E., Scott, D., Cassell, J.A.: Extracting information from the text of electronic medical records to improve case detection: a systematic review. J. Am. Med. Inform. Assoc. **23**(5), 1007–1015 (2016). https://doi.org/10.1093/jamia/ocv180
3. Ayala Solares, J.R., et al.: Deep learning for electronic health records: a comparative review of multiple deep neural architectures. J. Biomed. Inform. **101**, 103337 (2020). https://doi.org/10.1016/j.jbi.2019.103337
4. Oudah, M., Shaalan, K.: NERA 2.0: improving coverage and performance of rule-based named entity recognition for Arabic. Nat. Lang. Eng. **23**, 441–472 (2016). https://doi.org/10.1017/s1351324916000097
5. Uzuner, Ö., South, B.R., Shen, S., DuVall, S.L.: 2010 i2b2/VA challenge on concepts, assertions, and relations in clinical text. J. Am. Med. Inform. Assoc. **18**(5), 552–556 (2011). https://doi.org/10.1136/amiajnl-2011-000203
6. Uzuner, O., Luo, Y., Szolovits, P.: Evaluating the state-of-the-art in automatic de-identification. J. Am. Med. Inform. Assoc. **14**(5), 550–563 (2007). https://doi.org/10.1197/jamia.m2444

7. Spositto, O.M., Bossero, J.C., Moreno, E.J., Ledesma, V.A., Matteo, L.R.: Lexical analysis using regular expressions for information retrieval from a legal corpus. In: Pesado, P., Gil, G. (eds.) CACIC 2021. CCIS, vol. 1584, pp. 312–324. Springer, Cham (2022). https://doi.org/10.1007/978-3-031-05903-2_21
8. Zhang, S., Elhadad, N.: Unsupervised biomedical named entity recognition: experiments with clinical and biological texts. J. Biomed. Inform. **46**(6), 1088–1098 (2013). https://doi.org/10.1016/j.jbi.2013.08.004
9. Eftimov, T., Koroušić Seljak, B., Korošec, P.: A rule-based named-entity recognition method for knowledge extraction of evidence-based dietary recommendations. PLoS ONE **12** (2017). https://doi.org/10.1371/journal.pone.0179488
10. Mumtaz, R., Qadir, M.A.: CustNER: a rule-based named-entity recognizer with improved recall. Int. J. Semant. Web Inf. Syst. **16**(3), 110–127 (2020). https://doi.org/10.4018/ijswis.2020070107
11. Zhang, S., Shen, Y., Tan, Z., Wu, Y., Lu, W.: De-bias for generative extraction in unified NER task. In: Proceedings of the 60th Annual Meeting of the Association for Computational Linguistics, pp. 808–818. Association for Computational Linguistics, Dublin (2022). https://doi.org/10.18653/v1/2022.acl-long.59
12. Wu, K., et al.: Named entity recognition of rice genes and phenotypes based on BiGRU Neural Networks. Comput. Biol. Chem. **108**, 107977 (2024). https://doi.org/10.1016/j.compbiolchem.2023.107977
13. Alnazzawi, N., Thompson, P., Batista-Navarro, R., Ananiadou, S.: Using text mining techniques to extract phenotypic information from the PhenoCHF corpus. BMC Med. Inform. Decis. Mak. **15** (2015). https://doi.org/10.1186/1472-6947-15-s2-s3
14. Leaman, R., Wei, C.-H., Zou, C., Lu, Z.: Mining chemical patents with an ensemble of open systems. Database (Oxford) **2016** (2016). https://doi.org/10.1093/database/baw065
15. Bikel, D.M., Miller, S., Schwartz, R., Weischedel, R.: Nymble: a high-performance learning name-finder. In: Fifth Conference on Applied Natural Language Processing, pp. 194–201. Association for Computational Linguistics, Washington DC (1997). https://doi.org/10.3115/974557.974586
16. Lample, G., Ballesteros, M., Subramanian, S., Kawakami, K., Dyer, C.: Neural architectures for named entity recognition. In: Proceedings of the 2016 Conference of the North American Chapter of the Association for Computational Linguistics: Human Language Technologies, pp. 260–270. Association for Computational Linguistics, San Diego (2016). https://doi.org/10.18653/v1/n16-1030
17. Krishnan, V., Manning, C.D.: An effective two-stage model for exploiting non-local dependencies in named entity recognition. In: Proceedings of the 21st International Conference on Computational Linguistics and 44th Annual Meeting of the Association for Computational Linguistics. pp. 1121–1128. Association for Computational Linguistics, Sydney (2006). https://doi.org/10.3115/1220175.1220316
18. Ke, J., Wang, W., Chen, X., Gou, J., Gao, Y., Jin, S.: Medical entity recognition and knowledge map relationship analysis of Chinese EMRs based on improved BiLSTM-CRF. Comput. Electr. Eng. **108**, 108709 (2023). https://doi.org/10.1016/j.compeleceng.2023.108709
19. Lyu, C., Chen, B., Ren, Y., Ji, D.: Long short-term memory RNN for biomedical named entity recognition. BMC Bioinform. **18**, 462 (2017). https://doi.org/10.1186/s12859-017-1868-5
20. Gorinski, P.J., et al.: Named entity recognition for electronic health records: a comparison of rule-based and machine learning approaches. arXiv preprint arXiv:1903.03985 (2019)
21. Mir, T.H., et al.: Deidentification and temporal normalization of electronic health record notes using large language models: the SREDH/AI-Cup 2023 deidentification competition. In: 2024 International Workshop on Deidentification of Electronic Medical Record Notes. Springer, Kaohsiung (2024)

22. Jonnagaddala, J., Chen, A., Batongbacal, S., Nekkantti, C.: The OpenDeID corpus for patient de-identification. Sci. Rep. **11**, 19973 (2021). https://doi.org/10.1038/s41598-021-99554-9

23. Yan, X., Feng, S., Tang, Y., Yin, P., Deng, D.: Blockchain-based verifiable and dynamic multi-keyword ranked searchable encryption scheme in cloud computing. J. Inf. Secur. Appl. **71**, 103353 (2022). https://doi.org/10.1016/j.jisa.2022.103353

24. Minh, Q.T., Tan, D.P., Le Hoang, H.N., Nhat, M.N.: Effective traffic routing for urban transportation capacity and safety enhancement. IATSS Res. **46**(4), 574–585 (2022). https://doi.org/10.1016/j.iatssr.2022.10.001

25. Malik, H., Tian, Z.: A framework for collecting YouTube meta-data. Procedia Comput. Sci. **113**, 194–201 (2017). https://doi.org/10.1016/j.procs.2017.08.347

26. Powers, D.M.W.: Evaluation: from precision, recall and F-measure to ROC, informedness, markedness and correlation. arXiv preprint arXiv:2010.16061 (2020)

27. Chen, A., Jonnagaddala, J., Nekkantti, C., Liaw, S.T.: Generation of surrogates for de-identification of electronic health records. Stud. Health Technol. Inform. **21**, 70–73 (2019). https://doi.org/10.3233/SHTI190185

28. Alla, N.L., Chen, A., Batongbacal, S., Nekkantti, C., Dai, H.-J., Jonnagaddala, J.: Cohort selection for construction of a clinical natural language processing corpus. Comput. Methods Programs Biomed. Update **1**, 100024 (2021). https://doi.org/10.1016/j.cmpbup.2021.100024

29. Liu, J., et al.: OpenDeID pipeline for unstructured electronic health record text notes based on rules and transformers: deidentification algorithm development and validation study. J. Med. Internet Res. **25** (2023). https://doi.org/10.2196/48145

30. AI-Cup 2023 website. https://codalab.lisn.upsaclay.fr/competitions/15425

31. IW-DMRN website. https://www.sredhconsortium.org/sredh-competitions/sredhai-cup-2023/2024-iw-dmrn

32. SREDH Consortium website. https://www.sredhconsortium.org/

A Hybrid Approach to the Recognition of Sensitive Health Information: LLM and Regular Expressions

Tsai-Yuan Huang⑩, Jun-Fu Shih⑩, Yun-Chien Hsieh⑩, and Hui-Hsien Feng$^{(\boxtimes)}$⑩

Department of English, National Kaohsiung University of Science and Technology,
1 University Road, Yanchao District, Kaohsiung City 824005, Taiwan
{F112133104,F111133110,F111133104,hhfeng}@nkust.edu.tw

Abstract. In the era of digital technology, it is crucial to prioritize privacy protection in healthcare. While Electronic Health Records text notes (EHR text notes) are widely used, securely removing patients' private information from these records presents a significant challenge. This study investigates the utilization of a combination of large language models (LLM) and regular expression techniques to improve the extraction and standardization of Sensitive Health Information (SHI) from medical records sourced from Australian institutions. In response to the limitations identified in the initial Pythia language model, a hybrid approach was developed, incorporating both deep learning and regular expressions. The primary aim of the current study was to enhance the accuracy of SHI recognition in medical records. The study utilized 2294 records for training and 950 records for testing from the institutions, i.e., the University of New South Wales, the Lowy Cancer Research Centre, and the Health Science Alliance Biobank. The study employed PyTorch for model training and regular expressions for SHI identification. The results indicated that the hybrid approach highly improved the accuracy of SHI recognition, leading to more satisfactory and complete outcomes. This study contributes to the discussion on optimizing SHI recognition in healthcare datasets and emphasizes the synergy between deep learning and regular expressions. The findings have practical implications for refining healthcare information systems, highlighting the potential of combining advanced language models with rule-based approaches to enhance precision in medical data extraction and standardization.

Keywords: Large Language Model · Regular Expressions · Sensitive Health Information · Pythia

1 Introduction

The incorporation of large language models (LLMs) [1] in clinical healthcare is considered a critical direction for the future of intelligent healthcare. However, system and software developers often overlook privacy concerns when interacting with LLMs. This oversight could potentially lead to a crisis including legal issues involving the leakage of confidential information. Moreover, when training these large language models, the

© The Author(s), under exclusive license to Springer Nature Singapore Pte Ltd. 2025
J. Jonnagaddala et al. (Eds.): IW-DMRN 2024, CCIS 2148, pp. 134–147, 2025.
https://doi.org/10.1007/978-981-97-7966-6_10

inclusion of actual private information in the training data, such as personal names, phone numbers, and ID numbers, could lead to privacy breaches because of the memory and interactive capabilities of large language models. Concurrently, amid the ongoing digital transformation, healthcare institutions worldwide have increasingly adopted electronic health record systems, making electronic health records a primary source for medical data analysis. These text-based electronic medical records within medical institutions often contain private or confidential information about patients. For example, specific information in electronic medical records, such as the patient's birthdate, appointment time, and attending physician, can potentially lead to the discovery of the patient's identity. Although Electronic Health Records text notes (EHR text notes) contain valuable medical investigations, they cannot be openly shared, so the vast majority of medical investigators can only access de-identified notes to protect the confidentiality of patients [2]. Therefore, it is crucial to develop techniques for removing patient privacy information from textual content to advance medical research using electronic medical records.

De-identification of clinical notes is a critical technique to protect the privacy and confidentiality of patients [3]. It also could build patient trust in the use of medical records in research [4]. Therefore, the de-identification of EHR text notes has been approached with various methods, including natural language processing (NLP) techniques and deep learning methods. Many studies have investigated the use of deep learning techniques for de-identifying clinical notes. For example, Dernoncourt et al. [2] found that artificial neural networks (ANNs) show superior effectiveness compared to previously published systems for de-identifying EHR text notes, even without requiring manual feature engineering. However, applying deep learning methods to develop de-identification strategies poses several challenges. Norgeot et al. [5] emphasized the challenges of developing reliable de-identification methods because of the limited availability of notes and the varying nature of SHI across different hospitals and departments. Alla et al. [6] also indicate that the overall quality of electronic health record (EHR) data is low due to inconsistent clinical coding and metadata standards across hospitals. Therefore, it is crucial to identify and overcome these obstacles to create a high-quality database.

Moreover, the scarcity of training data can lead to overfitting in deep learning models, resulting in poor generalization. This limitation can affect the performance of de-identification models based on deep learning, especially in cross-institutional settings where the models must be customized using local clinical notes [3].

In the field of data privacy and de-identification, the use of deep learning and LLM is highly influential. LLMs, known for their strong named entity recognition (NER) capabilities [7], have demonstrated their effectiveness in accurately identifying sensitive information in documents, which is a crucial step in the de-identification process [8]. Roy and Mitra [9] designed an innovative LLM to automatically identify and conceal personal information, thereby ensuring the implementation of robust data privacy measures. The findings indicate that these LLMs can accurately identify sensitive data, facilitating seamless concealment at the application layer during data distribution.

While applying LLMs faces obstacles, such as limited availability of data, variability in SHI, and potential overfitting due to insufficient training data, rule-based approaches offer a complementary approach. Liu et al. [10] have successfully utilized a hybrid approach of rules and transformers for de-identification tasks, achieving significant results. Inspired by such successes, our approach also integrates a rule-based approach, i.e., regular expressions, with deep learning techniques to address the challenges associated with de-identification effectively.

A regular expression, commonly referred to as a regex, is a string of characters that specifies a particular search pattern. It is extensively utilized for identifying patterns within text strings and facilitating activities such as parsing, data retrieval, and search-and-replace operations [11]. Kleene initially introduced the concept of regular expressions in the 1950s as a component of his research on regular events and finite-state automata [12].

Regular expressions have been used in de-identification methods, especially in the field of clinical documentation. For example, An et al. [13] proposed a de-identification method using regular expression rules and pre-trained BERT, demonstrating the utility of regular expressions in modern de-identification methodologies. Furthermore, Meystre et al. [14] highlighted the effectiveness of machine-learning-based systems with regular expression template features for de-identifying SHI in textual documents. Regular expressions offer a rule-based and flexible solution [15]. Therefore, integrating with deep learning, regular expressions contribute to a more comprehensive and robust de-identification strategy, enhancing the overall performance and flexibility of the models. This collaborative approach harnesses the strengths of both regular expressions and an LLM to address the challenges posed by the complexities of clinical note data.

In response to the privacy concerns associated with the use of electronic medical records and large language models, the SREDH Consortium [16] organized the International Workshop on AI CUP 2023-Privacy Protection and Standardization of Electronic Medical Record Competition [17, 18]. This competition aims to address two goals:

1. Extraction of patient privacy information: Identifying and categorizing all instances of sensitive health information (SHI) mentioned in each electronic medical record.
2. Standardization of the temporal information: Standardizing different formats of time descriptions mentioned in the medical cases into a uniform standard.

The details about the competition and its objectives are available on the IW-DMRN website [19].

Our team, under the name and code TEAM_4482, participated in this competition to evaluate AI models for entity recognition of sensitive health information and entity recognition and normalization of temporal information. In this competition, for task 1, we ranked 42, and for task 2, we ranked 10. Our overall team rank was 24.

In the current study, a hybrid approach was implemented to achieve the designated objectives. While an LLM served as a practical framework for extracting complex patterns and features from the electronic medical records (EMR) text notes data, regular expression techniques were also employed to provide complementary assistance in the extraction of relevant information. The LLM utilized in this study is Pythia, which is well-known for its reliable data organization and model structure [20], which enables efficient processing of textual data, particularly in the identification and management of sensitive information within clinical notes.

2 Methodology

A hybrid approach aiming to successfully recognize and standardize Sensitive Health Information from the medical records provided by Australian organizations, i.e., University of New South Wales (UNSW), Lowy Cancer Research Centre and Health Science Alliance Biobank, was developed. We utilized the Pythia-70M language model. Later, we developed a program utilizing regular expressions to complement the recognition of the identifiers in medical records. The identifiers recognized by the Pythia model and the regular expression patterns were merged to generate improved results.

2.1 Datasets

The OpenDeID v2 corpus consists of 3244 medical records from which we include 2294 records for the training dataset. The dataset is provided by the University of New South Wales (UNSW), Lowy Cancer Research Centre, and Health Science Alliance Biobank [21].

The eight main categories and 30 sub-categories of Sensitive Health Information from the annotation guidelines of HSA Study SHI Corpus are listed below in Table 1. Additionally, a set of 950 medical records from the same organizations was used as a test set to examine the extraction and standardization of the hybrid approach.

Table 1. The main categories and sub-categories of Sensitive Health Information.

SHI Category	Sub-category
NAME	PATIENT, DOCTOR, USERNAME
PROFESSION	No sub-categories
LOCATION	ROOM, DEPARTMENT, HOSPITAL, ORGANIZATION, STREET, CITY, STATE, COUNTRY, ZIP, OTHER
AGE	No sub-categories
DATE	DATE, TIME, DURATION, SET
CONTACT	PHONE, FAX, EMAIL, URL, IPADDRESS
IDs	SOCIAL SECURITY NUMBER, MEDICAL RECORD NUMBER, HEALTH PLAN NUMBER, ACCOUNT NUMBER, LICENSE NUMBER, VEHICLE ID, DEVICE ID, BIOMETRIC ID, ID NUMBER
OTHER	No sub-categories

Two instruments were developed to complete the tasks: (1) a trained Pythia language model in Python and (2) a SHI retrieving program in C#.

2.2 The Pythia Language Model

Pythia is a collection of language models ranging from 70M to 12B parameters to facilitate scientific research [20]. The 70 million parameters model (Pythia-70m) from the

Pythia Scaling Suite was selected to generate a model specialized in Sensitive Health Information recognition. It is a language model based on the GPT-NeoX library. According to Biderman et al. [20], Pythia provides several advantages in addition to its public availability, i.e.,

1. Models span several orders of magnitude of model scale.
2. All models were trained on the same data in the same order.
3. The data and intermediate checkpoints are publicly available for study.

Additionally, Pythia can standardize various date and time formats into the ISO 8601 standard to avoid confusion and misunderstanding caused by different date and time representation methods. The parameters used in model training and identifier recognition are provided in Table 2.

The training and prediction code utilized in this study were sourced from the tutorial course provided by the competition organizers. Additionally, the Python package islab-opendeid, designed for de-identifying patients' SHI within the OpenDeID v2 corpus [21], was integrated into the model training process. Specifically, the OpenDeidBatchSampler and collate_batch_with_prompt_template methods from the islab-opendeid package were employed for batch sampling the medical records dataset and collating the data for use with the PyTorch data loader function.

To assess the effectiveness of additional training epochs in improving the precise recognition of SHI, the training epoch was extended from 10 to 20. The evaluation revealed that the model trained over 20 epochs outperformed the model trained over 10 epochs, with 13,488 correct entries compared to 13,328 correct entries. Moreover, the batch size for data prediction was raised from 32 to 64, as sufficient video random access memory resources were available for PyTorch, and the data volume for prediction was lower than that for model training. Consequently, the decision was made to utilize the model trained over 20 epochs with a batch size of 64 for data prediction throughout the competition.

Table 2. Parameter values used in model training and data prediction.

SHI Category	Original Value	Value
Learning rate	3e−5	*
Epochs	10	20
Step	3000	*
Batch size	8 (Training)/32 (Prediction)	*/32 (Prediction)

2.3 The SHI Retrieving Program

We developed a Sensitive Health Information retrieving program using regular expression patterns to identify SHIs in the medical records to enhance recognition accuracy.

The regular expression patterns we composed were based on our observation of each medical record. They were tested through the website Regex 101 [22] and the

SHI retrieving program to verify their accuracy. The frequent patterns in the records are provided in Table 3.

The programming language C# and the platform .NET 8.0 were used to develop the pattern-based SHI retrieving program because of its high efficiency and minimal memory usage due to the availability of using Native AOT to compile into native code. The program only utilized the console window for input and output, which provided high portability on other operating systems, such as MacOS and Ubuntu. The patterns were hard coded into the code using the GeneratedRegexAttribute syntax as suggested by the compiler.

Besides the feature of retrieving SHI, the program also can merge the results of identifiers recognized by the trained model and generate a report. Furthermore, the program can convert the temporal information to be standardized into the ISO 8601 format. As shown in Table 4, each element is separated by a tab character (\t), and the names of identifiers are from the annotation guidelines of HSA Study SHI Corpus.

Table 3. The frequently appearing patterns of regular expression in medical records.

Identifier	Pattern
IDNUM	\d{2} [A-Z]\d{5,7}[A-Z0-9]?
MEDICALRECORD	\d{5,7}.[A-Z]{3}
PATIENT	^[A-Za-z]{2,}\s?[A-Za-z]{2,},\s?[A-Za-z]{2,}\s?[A-Za-z]+
DATE & TIME	?:(\d{3,4})Hrs\s{1}on\s)?(\d{1,2}[/\.]\d{1,2}[/\.]\d{2,4})(?:\s{1}at\s{1}(\d{1,2}:\d{2}))?
HOSPITAL	^Location:\s{2}((?:\d\/\d\s)?\w+\s?\w+\s?\w+\s?)\-\s?(\w+\s?\w+.?\w+)

Table 4. The format and example of the generated result.

Filename	Identifier	Start Index/End Index	Value	Standardized Value
10	TIME	151/170	28/08/2013 at 08:26	2013-08-28T08:26

2.4 Procedure

After gathering the data provided by the Australian organizations, three steps were taken. First, the model training process included loading training datasets ($n = 2294$), generating annotation files, and saving them into *.tsv files, using PyTorch to train the model with the dataset for 20 epochs, and saving models based on the values of average and minimum loss. Second, the SHI identification process included loading the trained model generated from the model training process, loading the datasets for identifier recognition, examining the identified value to see whether it needed to be standardized in ISO 8601 format, and writing all identified results into a text file. The flowchart of model training and data prediction process is provided in Fig. 1.

The last step is to process the output from the trained model. The output merge process included using regular expression patterns to recognize identifiers from the

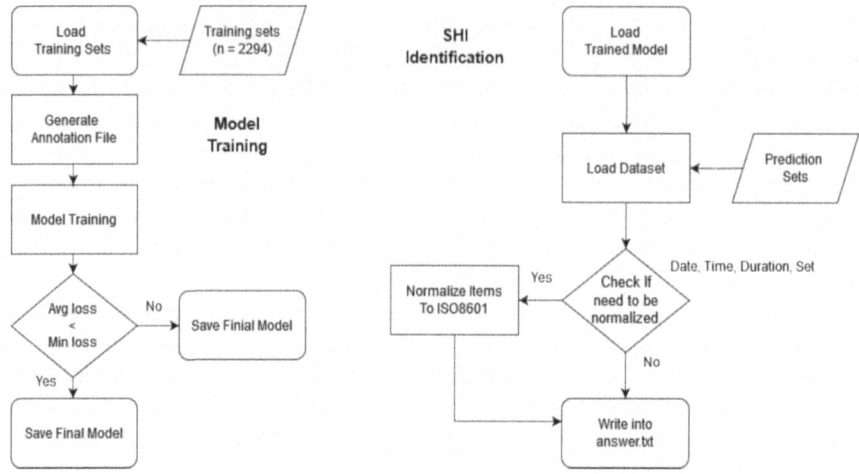

Fig. 1. The Sensitive Health Information identification process

medical records of the test set (hereafter the regex results), examining the identified value to see whether it needs to be standardized using ISO 8601, storing the identified regex results into a temporary file, comparing the results from the trained model for duplicates with the regex results and merging both results. Finally, the program generates the final results using the format shown in Table 4. The flowchart for merging outputs is provided in Fig. 2.

3 Results

3.1 Extraction of Patient Privacy Information

We trained the Pythia language model and utilized the SHI retrieving program to capture patient privacy data for the dataset of medical records. Table 5 illustrates the model performance of the Pythia language model, the SHI retrieving program, and the combination of both.

As shown in Table 5, the accuracy of the Pythia language model is 51.27%. Moreover, the accuracy of the SHI retrieving program is 75.47%, which is higher than that of the Pythia language model. Additionally, the combination of the Pythia language model and the SHI retrieving program was also conducted, and the accuracy was 69.77%. Compared to the sole use of the Pythia language model, the accuracy of the combination was higher. Moreover, a set of 950 medical records was used to test the extraction accuracy. The accuracy increased to 76.74%, which was also higher than that of the Pythia language model. Thus, the SHI retrieving program can enhance the performance of the Pythia language model on the extraction of patient privacy information. As the following example of data 2106:

Figure 3 shows the results identified by the Pythia language model. Nine entries were identified. On the contrary, Fig. 4 shows that the SHI retrieving program identified 16 entries. It can be seen that the Pythia language model captured fewer entries. Moreover,

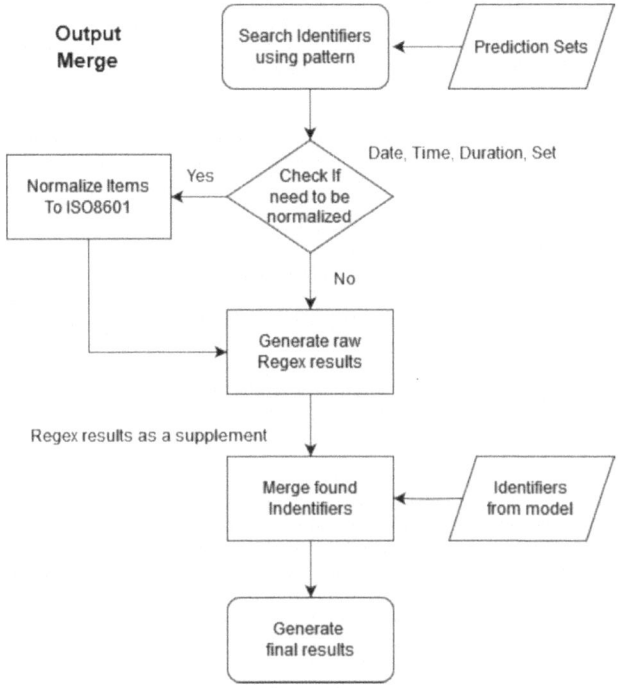

Fig. 2. Flowchart of recognizing identifiers using regex and merging the results.

Table 5. The extraction performance of the three models

Model	Accuracy
Pythia language model	51.27%
SHI retrieving program	75.47%
Pythia + Regex	69.77%
Pythia + Regex (test set)	76.74%

Fig. 5 demonstrates the merged results from Pythia + Regex in a total of 18 entries, which means the SHI retrieving program captures entries that the Pythia language model failed to capture.

3.2 Standardization of Temporal Information

The Pythia language model and the SHI retrieving program were utilized to standardize the temporal information for two medical record datasets. Table 6 shows the accuracy of temporal standardization trained by the Pythia language model, which is 54.31%. Then, the accuracy of standardization retrieved by the SHI retrieving program is 71.43%, which is higher than that of the Pythia language model. When Pythia + Regex was utilized, an

```
2106      IDNUM    14        24        83T2060400
2106      MEDICALRECORD      25        36        8372060.MAO
2106      PATIENT 38         60            REARY, Johnathan Jimmy
2106      IDNUM    71        79        83T20604
2106      STREET   80        88        Mortlake
2106      DATE     138       146       4/3/1993           1993-03-04
2106      DATE     166       176       16/06/2065         2065-06-16
2106      DEPARTMENT         197       214    ACUTE STROKE UNIT
2106      DOCTOR   218       236       INGER Eugenio Sour
```

Fig. 3. Example of data retrieved by the Pythia language model.

```
2106      IDNUM    14        24        83T2060400
2106      IDNUM    71        79        83T20604
2106      MEDICALRECORD      25        36        8372060.MAO
2106      PATIENT 38         60        REARY, Johnathan Jimmy
2106      DOCTOR   218       236       INGER Eugenio Sour
2106      DOCTOR   1560      1569      W Callais
2106      DOCTOR   1573      1584      W Krumvieda
2106      DOCTOR   1513      1515      PD
2106      DOCTOR   1517      1519      AO
2106      STREET   80        88        Mortlake
2106      CITY     89        101       DODGES FERRY
2106      STATE    103       106       ACT
2106      ZIP      108       112       6514
2106      DATE     138       146       4/3/1993        1993-03-04
2106      DATE     1523      1530      17.6.65 2065-06-17
2106      TIME     166       185       16/06/2065 at 19:35      2065-06-16T19:35
```

Fig. 4. Example of data retrieved by the SHI retrieving program.

```
2106      IDNUM    14        24        83T2060400
2106      MEDICALRECORD      25        36        8372060.MAO
2106      PATIENT 38         60        REARY, Johnathan Jimmy
2106      IDNUM    71        79        83T20604
2106      DATE     138       146       4/3/1993        1993-03-04
2106      DEPARTMENT         197       214    ACUTE STROKE UNIT
2106      DOCTOR   218       236       INGER Eugenio Sour
2106      DOCTOR   1560      1569      W Callais
2106      DOCTOR   1573      1584      W Krumvieda
2106      DOCTOR   1513      1515      PD
2106      DOCTOR   1517      1519      AO
2106      STREET   80        88        Mortlake
2106      CITY     89        101       DODGES FERRY
2106      STATE    103       106       ACT
2106      ZIP      108       112       6514
2106      DATE     1523      1530      17.6.65 2065-06-17
2106      TIME     166       185       16/06/2065 at 19:35      2065-06-16T19:35
2106      DOCTOR   220       236       GER Eugenio Sour
```

Fig. 5. Example of data retrieved by Pythia + Regex.

accuracy of 71.14% was obtained. As for the test set, Pythia + Regex gained an accuracy of 66.88%, which is higher than that of the sole use of the Pythia language model.

Table 6. The standardization performance of the three models

Model	Accuracy
Pythia language model	54.31%
SHI retrieving program	71.43%
Pythia + Regex	71.14%
Pythia + Regex (test set)	66.88%

Figure 6 illustrates the standardization for temporal information from the Pythia language model with an example from data 2058. Five entries were obtained, and all belonged to one type of temporal information, DATE. On the other hand, Fig. 7 demonstrates that nine entries were extracted and belonged to three types of temporal information: DATE, TIME, and DURATION.

```
2058   DATE    153    163    20/11/2021    2021-11-20
2058   DATE    354    361    3.12.63       2063-12-03
2058   DATE    388    394    3.9.63 2063-09-03
2058   DATE    6129   6135   3.9.63 2063-09-03
2058   DATE    6506   6513   3.12.63       2063-12-03
```

Fig. 6. Example of temporal standardization from the Pythia language model

```
2058   DATE     153    163    20/11/2021    2021-11-20
2058   DATE     354    361    3.12.63       2063-12-03
2058   DATE     388    394    3.9.63 2063-09-03
2058   DATE     3574   3581   28.9.63       2063-09-28
2058   DATE     388    394    3.9.63 2063-09-03
2058   DATE     354    361    3.12.63       2063-12-03
2058   TIME     183    202    25/07/2013 at 13:58  2013-07-25T13:58
2058   DURATION        508    514    3 week P3W
2058   DURATION        549    555    10 day P10D
```

Fig. 7. Example of temporal standardization from the SHI retrieving program

Furthermore, the same nine entries were captured by Pythia + Regex, which is more complete than when the Pythia language model was solely used. Thus, the SHI retrieving program can complement the results that the Pythia language model missed, as shown in Fig. 8.

However, the researchers found that there was incorrect standardization of temporal information from the Pythia language model. As shown in Fig. 9, when defining date and month, the Pythia language model could not identify which number is the date or the month if the number is a unit digit. As shown in the example of data 1097, "7/9/2063", "7"

```
2058    DATE    153     163     20/11/2021     2021-11-20
2058    DATE    354     361     3.12.63        2063-12-03
2058    DATE    388     394     3.9.63 2063-09-03
2058    DATE    6129    6135    3.9.63 2063-09-03
2058    DATE    6506    6513    3.12.63        2063-12-03
2058    DATE    3574    3581    28.9.63        2063-09-28
2058    DATE    354     361     3.12.63        2063-12-03
2058    TIME    183     202     25/07/2013 at 13:58 2013-07-25T13:58
2058    DURATION        508     514     3 week P3W
```

Fig. 8. Example of temporal standardization from Pythia + Regex

is the date, and "9" is the month. However, the Pythia language model standardized "7" as the month and "9" as the date, which is incorrect standardization. Thus, the training of the language model for temporal standardization needs further improvement. Overall, for the standardization of temporal information, it can be seen that the SHI retrieving program enhances the results of the Pythia language model.

```
1097    DATE    60      68      7/9/2063       2063-07-09
| Correct => 7/9/2063   2063-09-07
1135    DATE    124     132     3/6/1989       1989-03-06
| Correct => 3/6/1989   1989-06-03
1250    DATE    3136    3142    2/7/63 2063-02-07
| Correct => 2/7/63     2063-07-02
1250    DATE    5006    5012    4/8/63 2063-04-08
| Correct => 4/8/63     2063-08-04
1444    DATE    337     345     8.7.2063       2063-08-07
| Correct => 8.7.2063   2063-07-08
```

Fig. 9. Example of incorrect time standardization for the Pythia language model

4   Discussion

This study aims to identify patient privacy information and standardize the temporal information of medical records by the Pythia language model and SHI retrieving program. Results suggest that the SHI retrieving program can complement the performance of the Pythia language model with the extraction of patient privacy information.

Moreover, the SHI retrieving program enhanced the results of the Pythia language model for the standardization of temporal information. Likewise, An et al. [12] and Meystre et al. [13], demonstrated the utility of regular expressions integrated with deep learning to enhance the overall performance of models for de-identification. Additionally, the present results were similar to Liu et al. [10], who presented a hybrid de-identification pipeline called OpenDeID, which integrated associative rules, supervised deep learning, and pre-trained language models. Their results showed that the OpenDeID pipeline achieved a significant performance and has been utilized at a large tertiary teaching hospital, which also provided that the hybrid approach improved the accuracy of SHI recognition.

However, although Liu et al. [10] utilized regular expressions to detect certain types of incorrect tokens in the pre- and post-processing, after tokenization, they still needed to consider different types of mismatched data. Thus, it is important to notice that this study utilized regular expressions to capture patient information in different categories to complement the performance of the Pythia language model in the extraction of patient privacy information. This research has pointed to the potential of a hybrid approach to improve the efficacy of identifying Sensitive Health Information from medical records.

5   Conclusion

This paper presented a hybrid approach to de-identify SHI in medical records by utilizing the Pythia language model and the SHI retrieving program. We leveraged the advantages of both programs to identify SHI as comprehensively as possible. Furthermore, while the Pythia language model plays a crucial role in identifying and standardizing medical records, we recognize the inherent constraints associated with solely relying on an LLM. Although LLMs are powerful, difficulties may be encountered in managing the complexities of labeling variations in medical data types. In light of this, we strategically enhanced our approach by incorporating regular expressions, a rule-based approach for capturing language patterns.

This hybrid approach leverages the semantic understanding of the Pythia language model and the precision of regular expression extractions, providing a solution for dataset capture and standardization within complex medical records. In future work, additional regular expression patterns can be defined for SHI categories with low accuracy. Moreover, the temporal standardization with LLMs needs improvement to enable the models to generate correct time standardization answers based on the original text.

Acknowledgments. We would like to thank the organizer of the AI CUP 2023 Autumn Competition held by the Artificial Intelligence Competition and Annotation Data Collection Project Office of the Ministry of Education, Taiwan. We would also like to extend our gratitude to the organizations that provided the medical records for analysis, i.e., the University of New South Wales (UNSW), Lowy Cancer Research Centre, and Health Science Alliance Biobank. Additionally, we recognize the IW-DMRN workshop (https://www.sredhconsortium.org/sredh-com petitions/sredhai-cup-2023/2024-iw-dmrn) and the SREDH Consortium (https://www.sredhcons ortium.org/) for their valuable contributions to our research endeavors.

References

1. Tokayev, K.-J.: Ethical implications of large language models: a multidimensional exploration of societal, economic, and technical concerns. Int. J. Soc. Anal. **8**(9), 17–33 (2023). https://norislab.com/index.php/ijsa/article/view/42
2. Dernoncourt, F., Lee, J.Y., Uzuner, O., Szolovits, P.: De-identification of patient notes with recurrent neural networks. J. Am. Med. Inform. Assoc. JAMIA **24**(3), 596–606 (2017). https://doi.org/10.1093/jamia/ocw156

3. Yang, X., Bian, J., Wu, Y.: Customize deep learning-based de-identification systems using local clinical notes - a study of sample size. medRxiv (2020). https://api.semanticscholar.org/CorpusID:221094470

4. Fernandes, A.C., et al.: Development and evaluation of a de-identification procedure for a case register sourced from mental health electronic records. BMC Med. Inform. Decis. Mak. **13**(7) (2013). https://doi.org/10.1186/1472-6947-13-71

5. Norgeot, B., Muenzen, K., Peterson, T.A., et al.: Protected Health Information filter (Philter): accurately and securely de-identifying free-text clinical notes. NPJ Digit. Med. **3**(57) (2020). https://doi.org/10.1038/s41746-020-0258-y

6. Alla, N.L.V., Chen, A., Batongbacal, S., Nekkantti, C., Dai, H.-J., Jonnagaddala, J.: Cohort selection for construction of a clinical natural language processing corpus. Comput. Methods Programs Biomed. Update **1**, 100024 (2021). https://doi.org/10.1016/j.cmpbup.2021.100024

7. Chen, A., Jonnagaddala, J., Nekkantti, C., Liaw, S.T.: Generation of surrogates for de-identification of electronic health records. Stud. Health Technol. Inform. **264**, 70–73 (2019). https://doi.org/10.3233/SHTI190185

8. Liu, Z., et al.: DeID-GPT: zero-shot medical text de-identification by GPT-4. arXiv preprint (2023). https://arxiv.org/abs/2303.11032

9. Roy, S., Mitra, M.: Identification and processing of PII data, applying deep learning models with improved accuracy and efficiency. J. Data Acquisit. Process. **34** (2018)

10. Liu, J., et al.: OpenDeID pipeline for unstructured electronic health record text notes based on rules and transformers: deidentification algorithm development and validation study. J. Med. Internet Res. **25**(e48145) (2023). https://doi.org/10.2196/48145

11. Spishak, E., Dietl, W., Ernst, M.D.: A type system for regular expressions. In: Proceedings of the 14th Workshop on Formal Techniques for Java-like Programs. FTfJP '12, pp. 20–26. Association for Computing Machinery, Beijing, China (2012). https://doi.org/10.1145/2318202.2318207

12. Brzozowski, J.A.: Derivatives of regular expressions. J. ACM **11**(4), 481–494 (1964). https://doi.org/10.1145/321239.321249

13. An, J., Kim, J., Sunwoo, L., Baek, H., Yoo, S., Seunggeun, L.: De-identification of clinical notes with pseudo-labeling using regular expression rules and pre-trained BERT (2023). https://doi.org/10.21203/rs.3.rs-2672115/v1

14. Meystre, S.M., Friedlin, F.J., South, B.R., et al.: Automatic de-identification of textual documents in the electronic health record: a review of recent research. BMC Med. Res. Methodol. **10**(70) (2010). https://doi.org/10.1186/1471-2288-10-70

15. Turchin, A., Kolatkar, N.S., Grant, R.W., Makhni, E.C., Pendergrass, M.L., Einbinder, J.S.: Using regular expressions to abstract blood pressure and treatment intensification information from the text of physician notes. J. Am. Med. Inform. Assoc. **13**(6), 691–695 (2006). https://doi.org/10.1197/jamia.M2078

16. IW-DMRN website. https://www.sredhconsortium.org/sredh-competitions/sredhai-cup-2023/2024-iw-dmrn

17. AI-Cup 2023 website. https://codalab.lisn.upsaclay.fr/competitions/15425

18. Mir, T.H., et al.: Deidentification and temporal normalization of electronic health record notes using large language models: the SREDH/AI-Cup 2023 deidentification competition. In: 2024 International Workshop on Deidentification of Electronic Medical Record Notes. Springer, Kaohsiung (2024)

19. SREDH Consortium website. https://www.sredhconsortium.org/

20. Biderman, S., et al.: Pythia: a suite for analyzing large language models across training and scaling. In: Krause, A., Brunskill, E., Cho, K., Engelhardt, B., Sabato, S., Scarlett, J. (eds.) Proceedings of the 40th International Conference on Machine Learning, vol. 202, pp. 2397–2430. PMLR (2023). https://proceedings.mlr.press/v202/biderman23a.html

21. Jonnagaddala, J., Chen, A., Batongbacal, S., et al.: The OpenDeID corpus for patient de-identification. Sci. Rep. **11**, 19973 (2021). https://doi.org/10.1038/s41598-021-99554-9

22. Regex 101 website. https://regex101.com

Patient Privacy Information Retrieval with Longformer and CRF, Followed by Rule-Based Time Information Normalization: A Dual-Approach Study

Fan-Pin Tseng[1,2](✉) [ID], Han-Chun Ko[1] [ID], Xiu-Yu Hou[1] [ID], Danang Wijaya[1] [ID], Connyn Kang-Lin Chang[1] [ID], and Richard Tzong-Han Tsai[1] [ID]

[1] National Central University, No. 300, Zhongda Road, Zhongli District, Taoyuan 320, Taiwan (R.O.C.)
ecyor@nari.org.tw, thtsai@g.ncu.edu.tw
[2] National Atomic Research Institute, No. 1000, Wenhua Road, Jiaan Village, Longtan District, Taoyuan 32546, Taiwan (R.O.C.)

Abstract. This study explores integrating the Longformer model with Conditional Random Fields (CRF) for enhancing Named Entity Recognition (NER) in the domain of healthcare data processing. It specifically focuses on patient privacy information retrieval and time information normalization, utilizing the comprehensive 'Artificial Intelligence CUP 2023: Privacy Protection and Medical Data Standardization Challenge Dataset'. This research is conducted within the context of the AI CUP 2023 competition, which is dedicated to the privacy protection and standardization of medical data. Our approach utilized the Longformer model, renowned for its effectiveness in handling extensive text sequences, and combined it with CRF to enhance entity recognition accuracy in Electronic Health Record (EHR) text notes. To tackle challenges such as lengthy texts and class distribution imbalances, we developed a specialized process for managing large-scale textual data. This involved segmenting extensive texts into manageable chunks of 4,096 characters, which allowed for more focused and efficient training. For prediction, we employed a sliding window technique to ensure seamless integration and analysis of these text segments. This strategy was crucial in accurately retrieving patient privacy information from lengthy healthcare records. Additionally, our methodology included the implementation of rule-based methods for time information normalization, further enhancing the applicability of our approach in the medical data domain. The combination of Longformer and CRF has proven effective in accurately identifying sensitive patient information and normalizing time-related data. This approach illustrates the synergy between deep learning models and traditional methods, showcasing a robust framework for enhancing the security and efficiency of healthcare data processing.

Keywords: Healthcare Named Entity Recognition · Longformer-CRF · EHR Text Notes Processing

J. Jonnagaddala et al. (Eds.): IW-DMRN 2024, CCIS 2148, pp. 148–161, 2025.
https://doi.org/10.1007/978-981-97-7966-6_11

1 Introduction

The integration of digital technologies in healthcare, particularly through the adoption of Electronic Health Record (EHR) text notes, has fundamentally transformed medical practice and research. EHR text notes serve as a vital infrastructure, facilitate the storage of extensive patient data, thereby supporting clinical decision-making and medical research. However, this digital transition brings forth significant challenges, especially in maintaining patient privacy, which calls for advanced solutions to protect sensitive information.

With the advent of artificial intelligence (AI) technologies, particularly large language models (LLMs) like OpenAI's ChatGPT, we are witnessing groundbreaking enhancements in digital health. These technologies have become indispensable in clinical settings, improving the efficiency and effectiveness of medical data analysis and processing. For example, ChatGPT's capabilities in generating differential diagnosis lists and aiding in the creation of various clinical documents highlight the potential of AI to bolster healthcare providers' productivity and precision [1–4].

Nevertheless, implementing AI in healthcare requires careful consideration of legal and ethical frameworks, including compliance with the Health Insurance Portability and Accountability Act (HIPAA) and the General Data Protection Regulation (GDPR). These regulations set forth strict standards for data privacy, necessitating AI solutions to employ robust de-identification techniques [5–7]. Advanced Named Entity Recognition (NER) technologies and transformer models, including those specifically designed for long texts such as the Clinical-Longformer, have become crucial in enhancing the anonymization of datasets that may still inadvertently contain identifiable information [8–11]. Several NER systems use a combination of multiple approaches [12, 13]. An innovative system utilized an ensemble of transformer models such as BERT and RoBERTa for Multilingual Complex Named Entity Recognition (MultiCoNER), surpassing individual model capabilities. This method achieved a significant 2.85% increase in F1 score over the best single model for the code-mixed task [14]. These models have shown superior performance in processing extensive clinical texts within EHR text notes, thereby providing effective solutions for identifying and protecting Sensitive Health Information (SHI).

In the context of our commitment to addressing the challenges of data privacy and the ethical use of AI in healthcare, our participation in the AI-CUP 2023 has become particularly relevant [15, 16]. This prestigious competition, organized by the Taiwan Ministry of Education in collaboration with the Intelligent Systems Laboratory at the National Kaohsiung University of Science and Technology, the Department of Bioinformatics and Medical Engineering at Asia University, and the SREDH Consortium, focuses on the de-identification of EHRs text notes to ensure patient data privacy [17, 18]. It promotes academic exchanges and fosters innovations in artificial intelligence, offering a platform for showcasing solutions that comply with privacy regulations like HIPAA. This effort underscores our alignment with global efforts to safeguard SHI while advancing AI research [22].

Our research, inspired by the goals of AI-CUP 2023, focuses on developing innovative AI methods for Patient Privacy Information Retrieval (PPIR) and Time Information Normalization (TIN). By addressing these key areas, we aim to contribute significantly to the ethical utilization of patient data in healthcare. Our involvement not only aligns with the competition's objectives of advancing AI applications in healthcare privacy but also underscores the practical implications of our study in privacy protection and data standardization within the healthcare sector.

2 Methods and Approach

2.1 Dataset Description

The dataset utilized in this study originates from the "AI CUP 2023 Autumn Competition Privacy Protection and Medical Data Standardization Competition: Decoding Clinical Cases and Let Data Tell Stories" [19, 20]. The OpenDeID v2 corpus consists of 3,244 EHR text notes in.txt format, systematically divided into a training set with 1,733 records and a validation set comprising 560 records. Designed to facilitate tasks such as PPIR, akin to NER, and TIN, the dataset encompasses 21 unique class categories. The most extensive text within this collection reaches 18,483 characters in length. A notable aspect of this dataset is the imbalance in class distribution, as depicted in Fig. 1. Training and validation datasets have an imbalance in class distribution. For example, in the training data, there are 'SET,' 'URL,' and 'ROOM' classes, but in the validation data, these classes are not found. The frequencies of each class are unbalanced, in the training data there are 6 classes whose frequencies are very small when compared to the DOCTOR class which has a lot of frequencies.

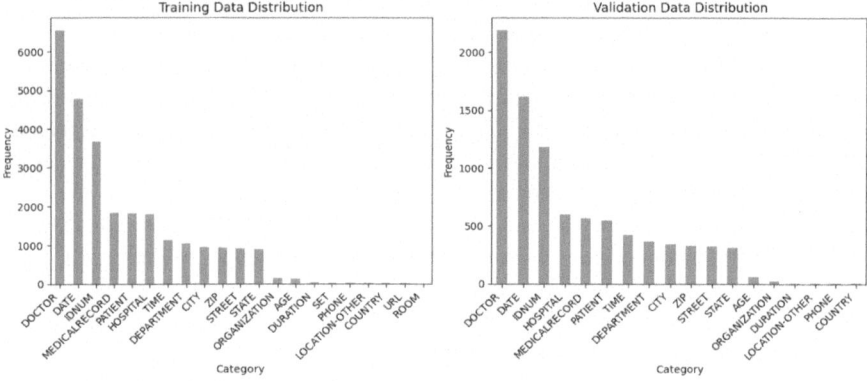

Fig. 1. Distribution of categories in training and validation datasets.

2.2 Pre-processing

We perform data cleaning by ensuring that the labels of EHR text notes are consistent with their corresponding text in the training and validation data, then replace them with the correct labels. Figure 2, which illustrates the distribution of text lengths in the "AI CUP 2023 competition" dataset, reveals that the majority of texts range between

2,000 and 5,000 characters. To tackle the challenges associated with processing such lengthy sequences, we selected the Longformer-base-4096 model, a choice driven by its capacity to manage extensive texts efficiently. Acknowledging the Longformer's limitation of handling a maximum of 4,096 tokens, we segmented longer texts into smaller portions. Each segment was meticulously capped at 4,096 characters, ensuring that every piece remained within the model's processing threshold. This segmentation was done with careful consideration, adjusting labels for each segment to maintain context and guarantee the accuracy of our training data.

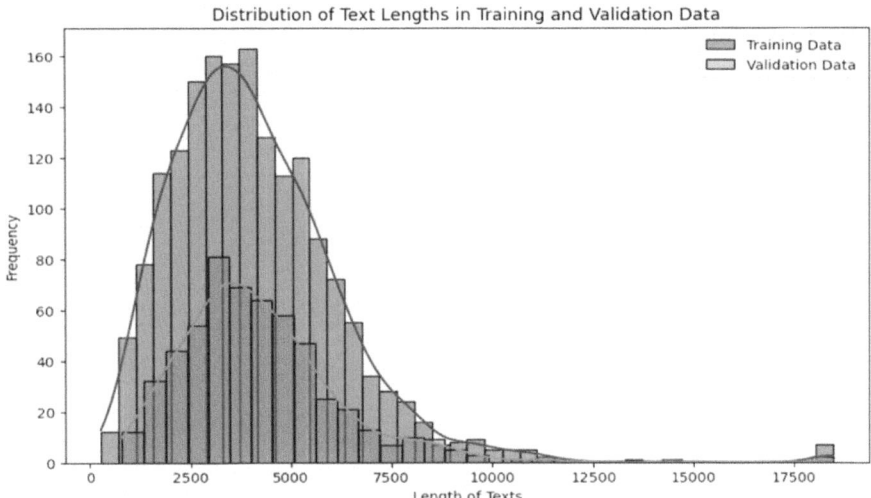

Fig. 2. Distribution of text lengths in "AI CUP 2023 competition" dataset.

2.3 Model Architecture

For the PPIR task of the AI CUP 2023 competition, we employed the Longformer model integrated with a CRF layer, as illustrated in Fig. 3. The CRF layer is instrumental in considering the context and inter-dependencies between labels in sequence prediction tasks, making it highly suitable for complex labeling tasks in medical texts. In our experimental setup, a batch size of 4 is selected to allow for more frequent model updates, which is crucial in handling the data's complexity and expected to improve generalization. We have set the learning rate at 1e-5 and the number of epochs at 20. AdamW is chosen as our optimizer, ideal for handling large pre-trained models such as Longformer-base-4096 due to its efficient management of sparse gradients and adaptive learning rate adjustments. To address the issues of class imbalance and data noise, the Macro-F1 score is employed as our primary evaluation metric. It calculates the F1 score for each class independently and averages them, ensuring a balanced evaluation across all classes and effectively counteracting the imbalance.

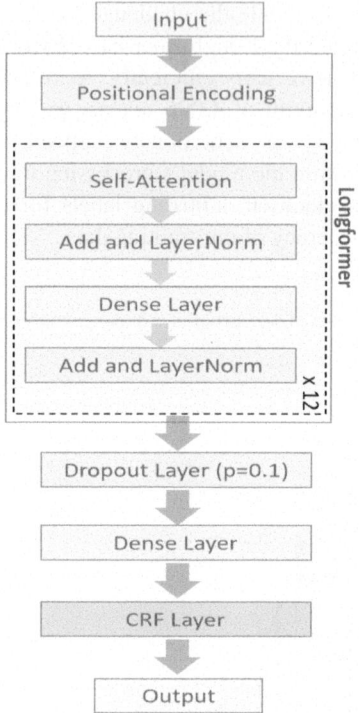

Fig. 3. Model architecture.

2.4 Approach Employed

As we transitioned to the prediction phase, highlighted in Fig. 4, we adopted a sliding window technique to effectively manage overlapping EHR text note segments. This method is crucial for maintaining continuity across the segmented texts, ensuring that no critical information is lost at the segment boundaries. The sliding window approach allows for the seamless integration of these segments, enhancing the model's capability to interpret and predict accurately across longer data sequences. Furthermore, to address the potential issue of information loss at these boundaries, we developed a strategy to merge predictions from overlapping segments by assigning a unified label. This was particularly applied when two predictions shared a label and overlapped in range, thus ensuring a coherent and comprehensive prediction output across the segmented texts.

In our study, we employed an ensemble learning technique across five distinct models, each trained on different datasets. The integration of these models was facilitated through a voting system. This system aggregates predictions from all models for each entity, identified by its label and start-end positions in the text. The final prediction for each entity is determined by selecting the most commonly predicted value across the models. This method enhances the robustness and accuracy of our entity recognition by leveraging the collective insights of multiple models.

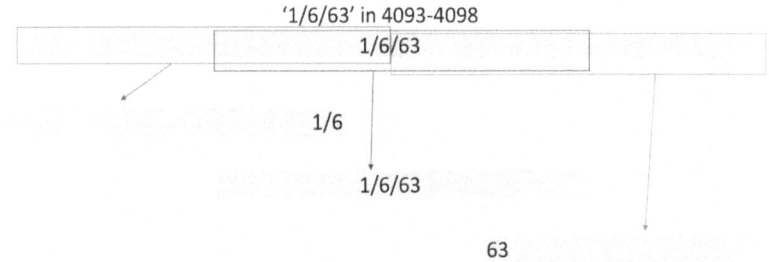

Fig. 4. Prediction phase sliding window.

After the implementation of the Longformer and CRF framework, we laid the ground-work for addressing the TIN task, which was further enhanced by a rule-based approach tailored to tackle the specific challenges presented by our dataset. This dual-strategy app-roach was particularly effective in managing the limited and unstructured nature of time-related data, such as 'DURATION', 'SET', 'DATE', and 'TIME' categories. The scarcity of 'DURATION' and 'SET' categories in the dataset—28 records for 'DURATION' and 14 for 'SET' in the training set, with notably fewer in the validation set—demanded a nuanced normalization strategy. Techniques like Word-to-Number conversion for textual numbers and Regex Matching were applied to identify and normalize duration patterns. Instances not conforming to standard patterns, such as 'several years' or '12/18', were marked as <unknown> to maintain data integrity.

For 'SET' data, 'SET' rule mapping was introduced to standardize textual expres-sions into a consistent format, allowing for accurate rule application even with mini-mal instances. The normalization process for 'DATE' and 'TIME' data also required addressing inconsistencies in formats and errors. Regular expressions were utilized to transform various date formats and correct common errors, including misplaced separa-tors or incomplete dates, standardizing the information into a consistent format. 'TIME' data normalization addressed inconsistencies and label errors, such as incorrect years and missing time parts, through regex-based transformations. This allowed for the sepa-ration of datetime data into 'DATE' and 'TIME' components for individual processing. The final step involved correcting errors in the separated data, such as incorrect AM/PM usage and misplaced separators, ensuring a high level of accuracy and consistency in our dataset's normalization.

3 Results

As depicted in Fig. 5, the model demonstrates a prominent diagonal line in the training set matrix, indicative of high true positive rates across most categories of EHR text notes. This pattern suggests effective learning of the training data. However, notable misclas-sifications are observed, particularly between 'DOCTOR' and 'PATIENT,' possibly due to similarities in their contextual mentions within EHR text notes.

In contrast, Fig. 6 shows reduced true positive rates overall, a common phenomenon as models generally perform better on trained data than on new data. The performance drop in categories such as 'DOCTOR' and 'DATE' in EHR text notes is significant,

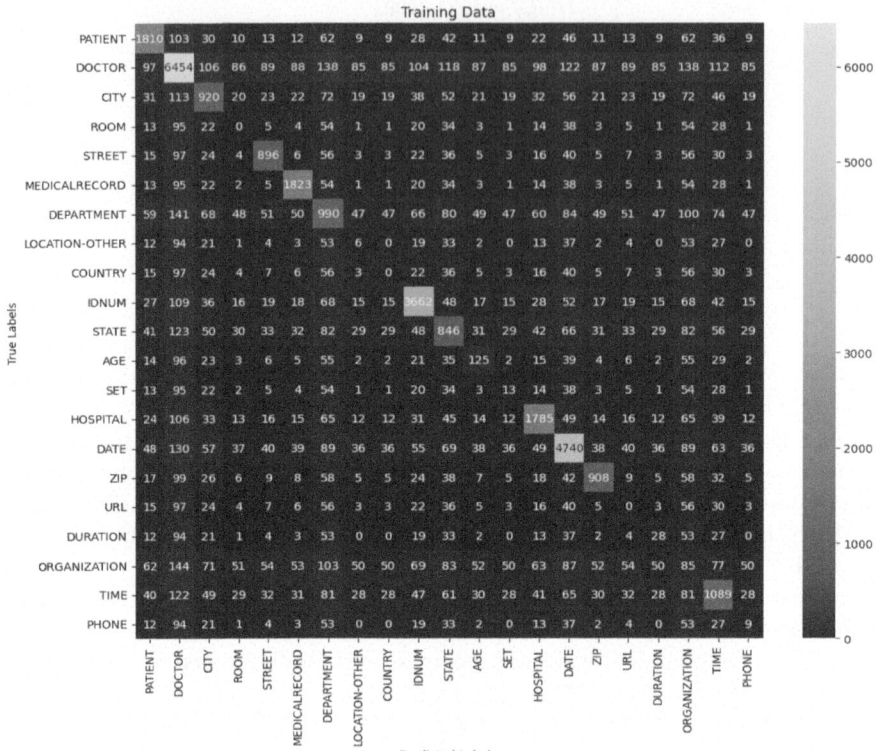

Fig. 5. The Longformer-base-4096 + CRF training data heatmap.

hinting at potential overfitting or challenges in generalizing from training to unseen data. Comparing heatmaps of training and validation data reveals potential overfitting, marked by robust performance on the training set and decreased performance on the validation set. This issue is especially pronounced in categories with limited data, like 'SET' and 'DURATION,' where the model seems to have memorized training examples instead of learning generalizable patterns. Our model faces challenges with data imbalance and noisy data. For example, the 'SET' category, having scant examples in the training set and none in the validation set, poses evaluation difficulties. Categories with more data, such as 'PATIENT' and 'MEDICALRECORD,' maintain comparatively better performance in the validation set than those with sparse data. Regarding noisy data, issues like mislabeling, exemplified by the incorrect time in ID:870, could lead the model to learn incorrect patterns. This mislearning can reduce performance on the validation set, where correct label distributions may differ from the training set's noisy data.

Based on the performance metrics illustrated in Fig. 7, we observed that our model's F1 score reaches its peak between epochs 15 and 17. To capitalize on this optimal performance, we have chosen the model from epoch 15 for a more in-depth analysis. This phase involves a comparative evaluation of the Longformer + CRF model against the

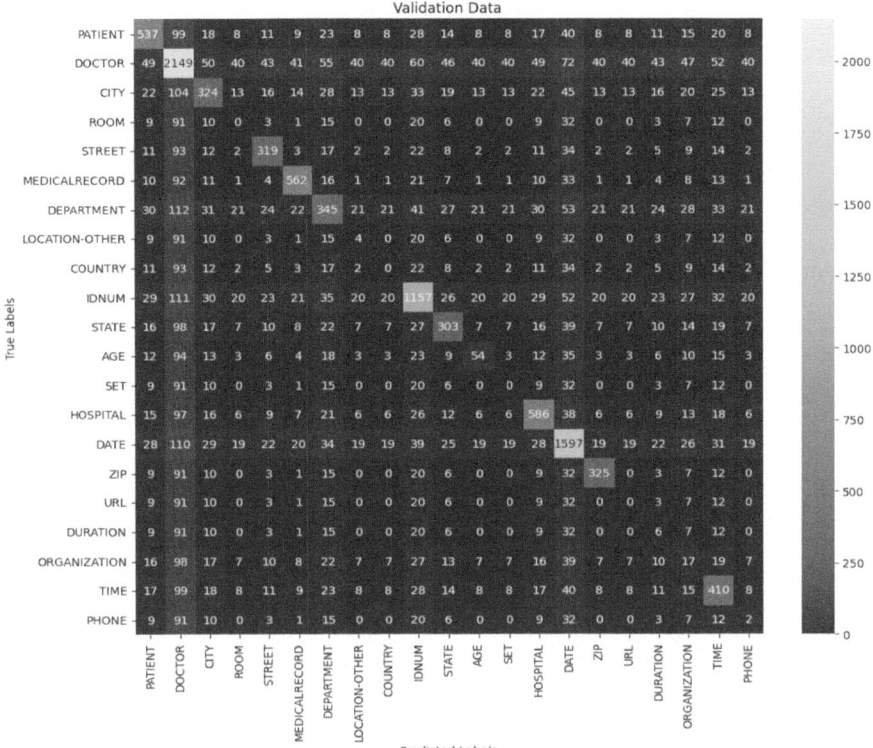

Fig. 6. The Longformer-base-4096 + CRF validation data heatmap.

standalone Longformer model. Our focus is on assessing their performance in various categories. This comparative analysis is vital for understanding the added value of integrating a CRF layer in complex NER challenges.

However, the Longformer-base-4096 + CRF model generally exhibits an improved F1 score, reflecting a balanced enhancement in both precision and recall. The incorporation of the CRF layer has made significant improvements in specific categories such as 'ORGANIZATION' and 'DURATION.' This improvement suggests that the CRF layer is more effective in capturing dependencies between labels, thereby enhancing the model's ability to accurately predict sequences of entities. Despite these overall enhancements, both models encounter challenges with the 'SET' and 'DURATION' categories. While the CRF layer offers some improvement, these categories still underperform compared to others, likely due to limited training data availability. The Longformer-base-4096 + CRF model demonstrates more consistent performance across various categories when compared to the standard Longformer-base-4096 model, which shows discrepancies between precision and recall in certain categories.

Table 1 presents a comprehensive comparison between the Longformer and Longformer + CRF models across different metrics. The data reveals that integrating a CRF layer with the Longformer model generally enhances performance, particularly in the

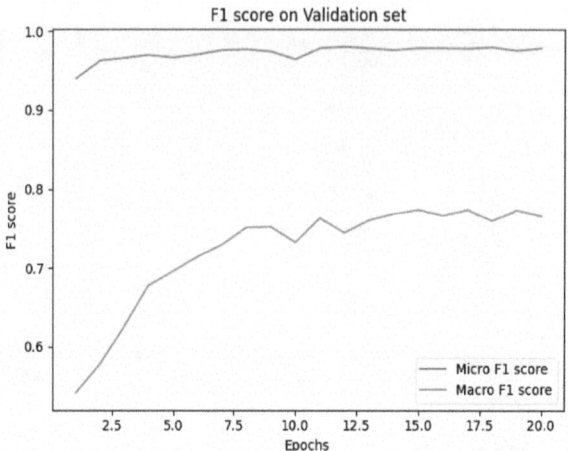

Fig. 7. Validation set performance across epochs for Longformer-base-4096 + CRF model.

validation set. This improvement is more pronounced in macro-level measurements, which are crucial for reflecting the performance across various categories, especially those less represented in the dataset. The table shows that for the 'Longformer + CRF' model, there is a notable improvement in macro precision, recall, and F1 score on the validation set, with values of 0.7684, 0.7828, and 0.7739, respectively. This enhancement indicates a more balanced performance across all categories, addressing the challenge of class imbalance effectively. In contrast, while the standalone Longformer model exhibits high micro-level scores (precision: 0.9755, recall: 0.9820, F1 score: 0.9788 on the validation set), its macro-level performance is lower compared to the Longformer + CRF model. This difference underscores the importance of the CRF layer in achieving a more uniform and equitable model performance, particularly in scenarios where class distribution is skewed.

To enhance our model's performance in the "AI CUP 2023" dataset, we employed an ensemble learning technique by combining five models from the 15th training epoch, using a voting system to amalgamate their predictions. This approach was designed to improve the consistency and reliability of our predictions across all categories, particularly for those with significant variations in F1 score as identified in our analysis. Our analysis, illustrated through a box plot comparison of F1 score on the validation set (Fig. 8), revealed that while certain categories like 'ROOM', 'COUNTRY', and 'URL' posed substantial challenges due to minimal learning, others such as 'SET', 'DURATION', 'PHONE', and 'LOCATION-OTHER' exhibited large score variations, likely due to their sparse representation in the dataset. This strategic ensemble method effectively stabilized performance, showcasing the power of model integration to overcome individual model weaknesses.

In the "AI CUP 2023" dataset, the limited presence of categories such as 'ROOM' and 'URL', with 'ROOM' only appearing once, significantly hampered the model's learning capability. The inability to recognize a specific reference like "Level 4 Campus Centre" reflects the model's struggle with minimal data exposure, underscoring the importance of

Table 1. Comparison of Longformer and Longformer + CRF Model.

Model	Data Type	Precision	Recall	F1 Score
Longformer + CRF	Train Micro	0.9858	0.9868	0.9863
	Validation Micro	0.9755	0.9823	0.9789
	Train Macro	0.8293	0.8273	0.8283
	Validation Macro	0.7684	0.7828	0.7739
Longformer	Train Micro	0.9877	0.9883	0.9880
	Validation Micro	0.9755	0.9820	0.9788
	Train Macro	0.8398	0.8422	0.8406
	Validation Macro	0.7393	0.7738	0.7529

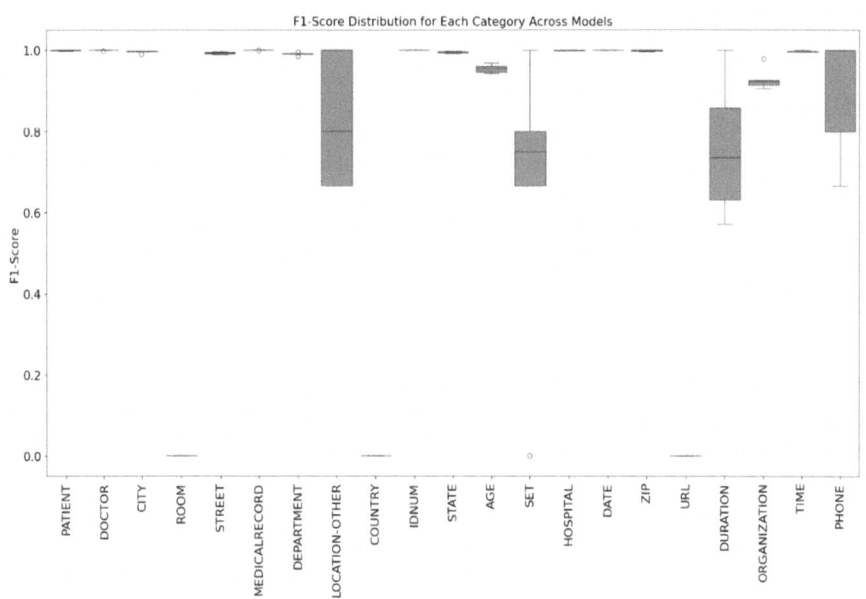

Fig. 8. Variability of F1 score across categories for ensemble model predictions on validation set.

a robust dataset for effective model training. The 'URL' category faced its own unique set of issues, with only three instances all tied to a singular data point, highlighting the model's limitations in segmenting text accurately despite recognizing patterns like "http." This scenario underlines the need for precision in both pattern recognition and data segmentation to facilitate learning from sparse data.

Overfitting was evident in the 'SET' category, which was predominantly represented by the term "twice" in its 14 training instances, leading to the model's memorization rather than understanding. Meanwhile, slightly larger datasets for 'PHONE' and 'COUN-TRY' categories showed that even a small increase in data could aid in feature learning, despite 'COUNTRY's struggle with feature capture due to its minimal instance count. 'LOCATION-OTHER' and 'DURATION' similarly displayed the impact of instance scarcity, with 'LOCATION-OTHER' benefiting from pattern memorization and 'DURATION' facing over-prediction issues.

Misclassifications and segmentation errors particularly affected categories with more common or specialized terminologies, such as 'ORGANIZATION' and 'HOSPITAL'. Correct detections were often undermined by improper segmentation, highlighting the intricate balance required between accurate identification and precise segmentation for successful entity recognition.

In the "AI CUP 2023" competition, our team, "TEAM_3879", applied our methodology, with details shared in our code repository at https://github.com/danangwijaya750/AI-CUP-2023-Fall. We achieved an F1 score of 0.8647 in the PPIR task, ranking us 6th, and scored 0.7259 in the TIN task, placing us 15th. These outcomes secured us an overall 7th place in the competition standings.

4 Discussion

In the "AI CUP 2023" competition, we faced significant challenges with unevenly distributed categories within the dataset of pathology reports, primarily consisting of EHR text notes. Some categories had very few examples, complicating our models' learning processes in those areas. Additionally, we encountered difficulties in recognizing specific names in the PPIR task. These names, not commonly used in typical NLP model training, presented tuning challenges for our models when presented in sparse examples. For the TIN task, standardizing time information proved challenging, especially when it was embedded within nouns or contextually related to surrounding text, making it difficult to normalize using regular expressions. To enhance performance, we adopted ensemble learning and rule-based approaches, significantly improving our models' capabilities to address these complex tasks.

This strategic adoption of ensemble learning techniques played a crucial role in enhancing our model's performance, notably in reducing prediction variability. It proved exceedingly beneficial for categories with sparse representation in the EHR text notes dataset, such as 'LOCATION-OTHER', 'DURATION', and 'SET'. By utilizing multiple models in an ensemble approach, each trained on distinct data splits, we substantially improved our model's learning from limited instances. Our analysis highlights the significant benefits of this technique, as demonstrated by the test set results. The categories with sparse representation achieved commendable F1 score: 'DURATION' with an F1 score of 0.89 from 12 instances, 'LOCATION-OTHER' at 0.75 from 6 instances, and 'SET' achieving 0.89 from just 5 instances. These results underscore the capacity of ensemble learning to stabilize and enhance accuracy for categories represented by minimal instances.

In our 'AI CUP 2023' study, we explored objectives aligned with those in the "Open-DeID Pipeline for Unstructured Electronic Health Record Text Notes Based on Rules and Transformers" study [21]. Both studies employed NLP pre-trained models combined with Conditional Random Fields (CRF) to improve NER performance for SHI.

Our approach, using Longformer with CRF, was particularly effective in processing long texts and elevating macro F1 score through ensemble learning for underrepresented categories. Our methodology achieved a micro F1 score of 0.9789 on the validation set, outperforming the OpenDeID study's score of 0.9659, which utilized BioBERT fine-tuned on a similar dataset. The exploration of various word embedding techniques in the "OpenDeID" study, including PMC, GloVe, and BioBERT, presents a promising avenue for further research. Emphasizing the benefits of employing diverse embeddings to improve model efficacy, we envisage significant potential in exploring a range of NLP models like Transformer-XL, BigBird, and XLNet. Their unique tokenization and processing capabilities could substantially enhance our methodology, addressing the complex predictive challenges encountered in healthcare NLP applications.

For future model enhancements in the "AI CUP 2023" competition, we aim to refine our strategy in several crucial areas. Developing predefined lists and adopting rule-based methods for static noun categories such as "country," "city," and "hospital" will capitalize on their inherent stability for more accurate classification. In the TIN task, addressing context-dependent terms like "Today," "Previous," and "Friday" necessitates sophisticated date conversion strategies. We plan to integrate Large Language Models (LLMs) for their advanced contextual comprehension and explore deep learning models trained to discern the nuances of date-related information in medical texts. Additionally, tackling the challenges posed by de-identification practices that alter date formats, like standardizing years to a specific value, is essential for improving date recognition and model generalization across various datasets. These enhancements are anticipated to significantly boost the model's performance and utility in healthcare NLP tasks.

5 Conclusion

In our study, we demonstrated the efficacy of the Longformer model in processing large, complex EHR text notes by segmenting them for more accurate predictions. We enhanced prediction accuracy by employing simple rules, regular expressions, and combining CRF with ensemble learning. Future improvements will focus on better handling of text edges during segmentation and enhancing the quality of our training data. Strategies such as overlapping training segments and incorporating more diverse data are anticipated to improve our model's learning capacity and adaptability across various scenarios.

We also found value in mixing rule-based methods with deep learning to process data more smartly, especially for medical texts. Adapting our methods to handle the different ways time and dates are expressed in various languages and cultures is important for making our model more flexible and accurate. Adding more context to our data, like using dates mentioned earlier in the text, can also help our model understand and interpret time information better. By focusing on these areas, we expect to improve our model's ability to handle medical texts more accurately and efficiently.

Acknowledgments. We thank the AI Cup 2023 organizers, particularly the Taiwan Ministry of Education, the Intelligent Systems Laboratory at National Kaohsiung University of Science and Technology, the Department of Bioinformatics and Medical Engineering at Asia University, and the SREDH Consortium, for their support and for providing the dataset crucial for our healthcare AI research.

Disclosure of Interests. The authors declare that they have no competing interests relevant to the content of this article to disclose.

References

1. Liu, J., Li, C., Liu, S.: Utility of ChatGPT in clinical practice. J. Med. Internet Res. **25** (2023)
2. Qiu, J., et al.: Large AI models in health informatics: applications, challenges, and the future. IEEE J. Biomed. Health Inform. **27**, 6074–6087 (2023)
3. Meskó, B., Topol, E.J.: The imperative for regulatory oversight of large language models (or generative AI) in healthcare. NPJ Digit. Med. **6**, 120 (2023)
4. Li, Y., Wehbe, R.M., Ahmad, F.S., Wang, H., Luo, Y.: A comparative study of pretrained language models for long clinical text. J. Am. Med. Inform. Assoc. **30**, 340–347 (2022)
5. Johnson, A.E.W., Bulgarelli, L., Pollard, T.J.: Deidentification of free-text medical records using pre-trained bidirectional transformers. In: CHIL '20, pp. 214–221. Association for Computing Machinery, New York, NY, USA (2020)
6. Meaney, C., Hakimpour, W., Kalia, S., Moineddin, R.: A comparative evaluation of transformer models for de-identification of clinical text data. arXiv:2204.07056 (2022)
7. Catelli, R., Gargiulo, F., Damiano, E., Esposito, M., De Pietro, G.: Clinical De-identification using sub-document analysis and ELECTRA. In: 2021 IEEE International Conference on Digital Health (ICDH), pp. 266–275 (2021)
8. Nadeau, D., Sekine, S.: A survey of named entity recognition and classification. Lingvisticæ Investigationes **30**, 3–26 (2007)
9. Keretna, S., Lim, C.P., Creighton, D.: A Hybrid Model for Named Entity Recognition Using Unstructured Medical Text, pp. 85–90 (2014)
10. Beltagy, I., Peters, M.E., Cohan, A.: Longformer: the long-document transformer. arXiv: 2004.05150 (2020)
11. Li, Y., Wehbe, R.M., Ahmad, F.S., Wang, H., Luo, Y.: Clinical-longformer and clinical-BigBird: transformers for long clinical sequences. arXiv:2201.11838 (2022)
12. Crichton, G.K.O., Pyysalo, S., Chiu, B., Korhonen, A.: A neural network multi-task learning approach to biomedical named entity recognition. BMC Bioinform. **18** (2017)
13. Knafou, J., Naderi, N., Copara, J., Teodoro, D., Ruch, P.: BiTeM at WNUT 2020 shared task-1: named entity recognition over wet lab protocols using an ensemble of contextual language models. In: Proceedings of the Workshop on Noisy User-Generated Text (W-NUT 2020), pp. 305–313 (2020)
14. Schneider, E.T., Zavala, R.M.R., Martínez, P., Moro, C., Paraiso, E.: UC3M-PUCPR at SemEval-2022 task 11: an ensemble method of transformer-based models for complex named entity recognition. In: Proceedings of the International Workshop on Semantic Evaluation (SemEval-2022), pp. 1448–1456 (2022)
15. SREDH/AI-Cup 2023: SREDH/AI-Cup 2023 Deidentification Competition. https://www.sredhconsortium.org/sredh-competitions/sredhai-cup-2023
16. Secure Research Environment for Digital Health (SREDH) Consortium. https://www.sredhconsortium.org/
17. Chen, A., Jonnagaddala, J., Nekkantti, C., Liaw, S.-T.: Generation of surrogates for de-identification of electronic health records. In: Studies in Health Technology and Informatics. MEDINFO 2019: Health and Wellbeing e-Networks for All, vol. 264, pp. 70–73. IOS Press (2019)
18. Jonnagaddala, J., Chen, A., Batongbacal, S., Nekkantti, C.: The OpenDeID corpus for patient de-identification. Sci. Rep. **11**, 19973 (2021)
19. CodaLab: Protection and Medical Data Standardization Competition: Decoding Clinical Cases, Letting Data Tell the Story. https://codalab.lisn.upsaclay.fr/competitions/15425

20. Alla, N.L.V., Chen, A., Batongbacal, S., Nekkantti, C., Dai, H.-J., Jonnagaddala, J.: Cohort selection for construction of a clinical natural language processing corpus. Comput. Methods Programs Biomed. Update **1**, 100024 (2021)
21. Liu, J., et al.: OpenDeID pipeline for unstructured electronic health record text notes based on rules and transformers: deidentification algorithm development and validation study. J. Med. Internet Res. (2023)
22. Mir, T.H., et al.: Deidentification and temporal normalisation of the electronic health record notes using large language models: the SREDH/AI-Cup 2023 deidentification competition. In: 2024 International Workshop on Deidentification of Electronic Medical Record Notes. Springer, Kaohsiung (2024)

A Deep Dive into the Application of Pythia for Enhancing Medical Information De-identification in the AI CUP 2023

Zhi-En Li[✉] 📵, Hong-Yang Zheng 📵, Kuan-Chieh Mao 📵, and Zhi-Wen Wei 📵

Lab 606, Department of Electrical Engineering, National Kaohsiung University of Science and Technology, 415 JianGong Road, Sanmin, Kaohsiung 807, Taiwan
julian43541572@gmail.com, {F112154129,F112154164}@nkust.edu.tw,
mkj900829@gmail.com

Abstract. This report systematically introduces the process and outcomes of participating in the AI CUP 2023 Privacy Protection and Electronic Medical Record Standardization competition. We employed various innovative methods, particularly opting for the use of the "EleutherAI/pythia-410m-deduped" large language model in the final stages. In terms of related work, the report cites previous research on GPT-3's role in summarizing medical literature, emphasizing the potential advantages and limitations of large language models in handling medical documents. Regarding algorithmic approaches and model architecture, the report delves into their text generation model, including its design, configuration, and training methods. Notably, the team conducts a comparative analysis of different models (70m, 410m, and 1b), examining their scores during training and at different stages. We extensively discuss innovative training methods, data processing, and batch processing, highlighting their efforts in model training, prediction, and post-processing. Lastly, the report provides a detailed analysis of the competition data, including the quantity and performance comparison of each category in the test set. In conclusion, the report discusses model performance, training periods, and comparisons between different models, pointing out potential directions for future improvements. The comprehensive and in-depth nature of the report underscores the team's efforts and innovations in the competition.

Keywords: Text Generation · Medical Literature · Model Training · Large Language Models · Model Performance

1 Introduction

In recent years, the rapid advancement of artificial intelligence, particularly large-scale language models, has revolutionized various fields, including clinical healthcare. These models hold immense potential for intelligent healthcare systems of the future. However, their widespread use raises concerns about privacy protection, particularly regarding the risk of sensitive information leakage from medical records. To address this challenge, efforts have been initiated to develop robust de-identification systems, spurring progress in medical research and smart medical applications.

© The Author(s), under exclusive license to Springer Nature Singapore Pte Ltd. 2025
J. Jonnagaddala et al. (Eds.): IW-DMRN 2024, CCIS 2148, pp. 162–176, 2025.
https://doi.org/10.1007/978-981-97-7966-6_12

The AI Cup 2023: Privacy Protection and Medical Data Standardization competition [6, 17] and the 2024 International Workshop on the de-identification of Electronic Medical Record Notes (IW-DMRN) were held in order to solve the aforementioned issues [7, 8]. Trust must be built, private health information must be de-identified, and the ethical and appropriate use of healthcare data for research analysis must be encouraged. Considering the significance of maintaining SHI privacy [16]. The AI Cup 2023 provides a forum for discussing problems pertaining to data protection and standards in medical records.

In our study, we draw inspiration from these influential works [3, 4] to develop a robust de-identification model. Building upon the Transformer model architecture, we utilize free-text clinical notes to achieve sequence labeling for de-identification purposes. Additionally, the methodologies for generating surrogates for de-identification, as highlighted in the work presented by Chen, Jonnagaddala [5], are crucial for enhancing the privacy of medical data. Insights from the recent de-identification competition [6–8], also inform our approach, demonstrating the effectiveness of large language models in practical de-identification tasks. We integrate preprocessing and postprocessing steps proposed in prior research to identify and replace sensitive information, ensuring the effectiveness of our model in real-world clinical settings. By incorporating state-of-the-art techniques, our research aims to enhance patient privacy protection and facilitate secure handling of Electronic Medical Record data by healthcare institutions.

2   Related Work

In the study presented by Jonnagaddala, Chen [1], automated methods have proven to hold tremendous potential. Previous studies have established corpora containing thousands of medical records, providing valuable reference data for annotating sensitive information. Furthermore, researchers continually strive to develop automated methods for handling sensitive information even within code-mixed EMRs [9, 10], with some methods even incorporating active learning to effectively reduce annotation workload. These studies provide crucial references for understanding the current state and future development of EMR de-identification technologies.

In the report presented by Shaib, Li [2], the application of GPT-3 in the context of medical literature summarization was explored. They specifically focus on single-document and multi-document summarization tasks and assess the factual accuracy of the summaries generated by GPT-3 through evaluations conducted by domain experts in the medical field. The study finds that GPT-3 performs well in single-document summarization but faces challenges in the context of multi-document summarization. This research underscores the potential and limitations of large LMs in handling medical literature, especially in aggregating multiple documents while maintaining factual accuracy. The development of resources like the SREDH database [7] and initiatives such as those described by the International Workshop on De-identification of electronic Medical Record Notes (IW-DMRN) [8] contribute significantly to the advancement of EMR de-identification technologies by fostering a collaborative environment for research and development.

In the study presented by Liu, Gupta [3], a combination of deep learning techniques and rule-based methods, including LSTM, CRFs, and pre-trained BERT models, is employed. This hybrid approach demonstrates superiority in handling de-identification in Electronic Medical Records, particularly when combined with rule-based methods using the Philter tool. The deep learning model in this report opts for minibatch gradient descent and the Adam optimizer, along with the utilization of pre-trained BERT models. The choice of optimizer, especially the use of the AdamW optimizer, has been proven crucial in enhancing learning speed and model performance.

Moore, Bulgarelli [4] focuses specifically on de-identification of patient data, utilizing Transformer neural network models, including BERT, RoBERTa, and DistilBERT, trained on publicly available data from i2b2 2014. The uniqueness of this paper lies in its comprehensive utilization of different Transformer models, further expanding the application scope in de-identification. Particularly noteworthy is the report's choice to use the HuggingFace transformers library, providing diverse model options, which significantly influences the model training process.

In the research presented by Jeong, Kim [11], they delved into the application of GPT-3 in the context of medical literature summarization. They specifically focus on both single-document and multi-document summarization tasks, evaluating the factual accuracy of summaries generated by GPT-3 through assessments from domain experts in the medical field. The study reveals that GPT-3 performs well in single-document summarization tasks but faces challenges in the context of multi-document summarization. This research underscores the potential of large LMs in processing medical literature, highlighting their capabilities and limitations, particularly in aggregating multiple documents while maintaining factual accuracy. Another work presented by ALLA, Aipeng [12], they not only delved into various aspects such as entity identification, extraction, document classification, and de-identification but also dedicate efforts to constructing comprehensive annotated corpora, especially in the application of long-term clinical reports, ensuring consistent annotation. Simultaneously, for specific medical domains like family history extraction and disease stage factor identification, previous works emphasize the importance of tailoring methods to address specific challenges [13].

3   Method

3.1   Dataset

The Dataset used for competition is OpenDeID v2 which comprises of 3,244 pathology reports. The OpenDeID v1 dataset serves as a crucial resource in this endeavor, comprising 2,100 pathology reports from 1,833 cancer patients across four urban hospitals in Australia. This publicly available Electronic Medical Record (EMR) corpus is meticulously annotated for protected health information entities, providing a standardized platform for evaluating de-identification technologies [1, 5, 12]. Its high quality and accuracy are ensured through meticulous annotation and consensus among annotators, facilitating research into patient privacy protection and supporting healthcare studies. Moreover, the groundbreaking work presented in the GPT3 paper [2] underscores the potential of large language models, particularly in medical summarization tasks. With its impressive accuracy in single-document summarization, GPT-3 with its 175 billion parameters serves as a solid foundation for our research. Leveraging this model's capabilities, we aim to advance deep natural language processing (NLP) tasks in the healthcare domain.

3.2 Experimental Setup

Our research environment is built on the Windows 11 operating system, utilizing Python 3.11.5 as the primary programming language. In terms of packages, we have integrated various powerful tools and libraries to support our language generation model research.

Firstly, we've incorporated the Transformers library, a crucial resource developed by the Hugging Face team. The library provides numerous pre-trained NLP models, such as BERT, GPT, RoBERTa, and supports various NLP tasks, including text classification and question-answering systems.

On another note, we've employed the Datasets library, also provided by Hugging Face. This library allows us to conveniently access and process a variety of NLP-related datasets, making it easier to load and handle the data required for our research.

In the realm of deep learning, we rely on PyTorch, the core library of the PyTorch deep learning framework, abbreviated as Torch. PyTorch provides the essential components for building and training models. To monitor the training process of our models effectively, we've adopted the TQDM progress bar library. This library enables us to display the progress of iterations clearly, providing better visualization and management. For parameter fine-tuning, we've chosen the PEFT (Parameter Efficient Fine-Tuning) method [14]. This method allows pre-trained LMs to adapt efficiently to various downstream applications without fine-tuning all parameters. Python's standard library also offers several useful modules, including Collections (collection data structures), Datetime (handling date and time data), IO (file I/O-related operations), and Re (regular expression operations).

Finally, the core pre-trained model for our research is "pythia-410m-deduped," developed by the EleutherAI team [15]. This large-scale LM is based on the GPT architecture. The model has undergone training on massive text datasets and has found widespread application in natural language processing tasks, including text generation, question-answering, and summarization.

3.3 Algorithmic Methods Exploration

Regarding algorithm methods and model architecture, the report delved into our text generation model, covering its design, configuration, and training methods. We highlighted innovations in model selection, data processing, and batch processing to underscore the efforts in model training, prediction, and post-processing. Finally, the report provided a detailed analysis of the competition data, including the quantity of each category in the test set and performance comparisons. In conclusion, we summarized our findings and delved into discussions on model performance, training epochs, and comparisons between different models.

It is worth noting that we also conducted a comparison of the 70m, 410m, and 1b models, analyzing their scores during training and the impact of different epochs.

This further enriched our comprehensive understanding of the challenges and strategies employed during the competition.

3.4   Model Identification Methodology

Dataset Processing. Data cleaning and preprocessing emerge as indispensable steps in the fields of data science and machine learning. These steps aim to ensure the quality of the dataset, thereby enhancing the performance and accuracy of the model.

The training data is sourced from the official training dataset released by the competition. After preprocessing, the data format (as shown in Table 1) consists of the medical record IDs, the sentence starting positions, sentences, and the corresponding answers.

Table 1. The format of the training data after preprocessing.

Medical records	Start positions	Sentence	Corresponding answers
10	1	Episode No: 09F016547J	IDNUM: 09F016547J
10	25	091016.NMT	MEDICALRECORD: 091016.NMT
10	37	SIZAR, HOWARD	PATIENT: SIZAR, HOWARD

We merged the provided two training subsets as the final training set to increase the quantity of training data. We then implemented customized batch processing methods to tailor the batch creation process aiming to enhance the efficiency of model training. Each training instance is converted into the following format before feeding it into the model for training: "Medical record sentence\n\n####\n\nAnswers corresponding to the sentence".

Training Procedure. The choice of using the "EleutherAI/pythia-410m-deduped" large LM as the base model is grounded in its robust capabilities and wide applicability. This decision is complemented by our unique training approach and dataset integration. We employed the AdamW optimizer as our training method. Additionally, the LoRA package from the peft toolkit is incorporated to reduce the number of training parameters and enhance training speed. In this configuration, the total training data comprises 406,120,448 samples, with 786,432 training samples included.

After training our model, the weight file is saved, eliminating the need to train from scratch for future predictions. When making predictions, the fine-tuned weights are loaded into the model. The test data is then input into the model to generate the predictions, which are post-processed to generate the desired output represented in a predefined format. Throughout the training phase, multiple experiments were conducted, with careful monitoring and recording of loss values during the training process. This highlights our close attention and in-depth analysis of the model training process. The meticulous adjustments made during training reflect our dedication to improving model performance.

Post-processing. Following the completion of predictions, we conducted additional post-processing and result analysis. This involved parsing and formatting the predicted results, as well as performing further processing. Such post-processing and analysis are crucial to ensure that the final predicted output aligns with the specific output requirements of the competition. This process not only enhances the interpretability of predictions but also ensures the practicality and accuracy of the model's output.

Table 2. The keyword settings for "age" and "duration".

AGE	DURATION
numbers+"yo," "yr," "yrs," "years old,"or "years-old" variants	numbers+"years," "yrs," "years ago,"
"age" with numbers	numbers+"months"
"in" with numbers+"s"	numbers+"weeks" or "wks"

Our analysis reveals writing patterns for specific SHI types. We use regular expressions (as shown in Table 2) to recognize AGE and DURATION-related SHIs and transform recognized SHIs according to the specified format. Based on the extracted string of a recognized SHI, we determine its starting position in the original sentence and calculate its end position based on the length of the string. The results are then stored in the output array. To ensure the quality of the output, the following rules are applied: (1) the output answer's string cannot exceed 50 characters, [2] the output is removed if the starting and ending positions are the same, and [3] the output is filtered out if the SHI type is not belonging to the desired SHI categories. In addition, due to the logic applied to AGE and DURATION, there may be cases where the predicted results are identical to the recognized results generated by the employed LM. To address this, we implement another post-processing to eliminate duplicate answers from the text file.

In addition to the 410m model, predictions are also made using 70m and 1b models. The decision to finally choose the 410m model for the competition is based on factors including the time constraints and result quality.

4   Results

4.1   Validation Data Sets

The competition provided two sets of the test data, consisting of a pre-competition test set (we refer it to as the validation set) and the official test set. The statistical information of the two dataset is shown in the Table 3.

4.2   Performance Comparison with and Without LoRA

From Table 4, although the utilization of LoRA appears to decrease the model's performance, leading to a decline in F-scores, this phenomenon may be attributed to the nature of the task and the functioning of LoRA. The primary function of LoRA involves introducing localized randomization during the model training process and subsequently aggregating the outcomes of this randomness. In certain scenarios, the compatibility between this localized randomness and the nature of the task may be inadequate, resulting in the model struggling to effectively learn or adapt during the optimization process.

Table 3. Statistics of the competition test set.

	Validation Set	Test Set
IDNUM	1,177	2,120
MEDICALRECORED	563	747
PATIENT	545	716
STREET	321	344
CITY	337	373
STATE	310	332
ZIP	325	353
DATE	1,612	2,459
DEPARTMENT	366	419
HOSPITAL	593	1,198
DOCTOR	2,191	3,327
TIME	418	470
AGE	57	51
DURATION	6	12
ORGANIZATION	24	74
PHONE	2	1
COUNTRY	2	0
LOCATTION-OTHER	4	6
SET	0	5
TOTAL	8,853	13,007

However, despite the observed decrease in scores, it is important to note that LoRA significantly reduces the training time. This time advantage, especially for complex and large-scale models, could be of paramount importance, enhancing the practicality and applicability of the model in real-world scenarios. Thus, even with a slight reduction in scores, the use of LoRA may offer advantages in terms of expeditiously completing model training within a relatively shorter timeframe, making it more suitable for real-world applications.

Table 4. Scores Comparison for 410m Model with and without LoRA (10 epochs).

Coding Type	With LoRA	Without LoRA
MEDICAL RECORD	0.8800945	0.865027
PATIENT	0.9527121	0.7893661
IDNUM	0.9298493	0.9336979
DATE	0.9405941	0.923832
DOCTOR	0.8742217	0.8313437
STREET	0.7709751	0.852071
CITY	0.9309392	0.946794
STATE	0.9553571	0.9701492
ZIP	0.972973	0.8776371
TIME	0.8365591	0.7165644
DEPARTMENT	0.7709923	0.823383
HOSPITAL	0.8119619	0.8477892
DURATION	0.75	0.6896552
AGE	0.8484849	0.8623853
ORGANIZATION	0.4697986	0.4588235
SET	0	0.5
LOCATION-OTHER	0	0.2857143
PHONE	0	0
Micro-avg.F	0.8906692	0.869721
Macro-avg.F	0.7088497	0.7534189
DATE	0.7921502	0.8021612
TIME	0.4004657	0.4730618
DURATION	0.7619048	0.8181818
SET	0	0.5714286
Micro-avg	0.7315078	0.7572296
Macro-avg	0.489716	0.6840467
Training Time	3:11:00	6:51:23

4.3 Performance Comparison with Different Epochs

From Table 5, it can be observed that the results at epochs = 10 are better than those at epochs = 5. However, with further increase in training epochs, the scores gradually decrease, indicating a possible overfitting issue. Therefore, we believe that in the case of epochs = 10, the model results for 410m are better. Hence, for other models of different sizes, epochs = 10 is used as the baseline for comparison.

Table 5. Scores Comparison for 410m Model with different epochs.

Coding Type	Epochs = 5	Epochs = 10	Epochs = 15	Epochs = 30
MEDICALRECORD	0.8783943	0.8800945	0.8753691	0.87877
PATIENT	0.8968254	0.9527121	0.9409369	0.9646569
IDNUM	0.9163237	0.9298493	0.9423761	0.9544376
DATE	0.9357231	0.9405941	0.9353995	0.9287577
DOCTOR	0.8681535	0.8742217	0.8947778	0.886424
STREET	0.832869	0.7709751	0.8938992	0.9034853
CITY	0.9381018	0.9309392	0.9360543	0.9381018
STATE	0.9237668	0.9553571	0.9569094	0.958209
ZIP	0.9602273	0.972973	0.9546743	0.945869
TIME	0.4068157	0.8365591	0.7975078	0.4877006
DEPARTMENT	0.7897436	0.7709923	0.8035488	0.7898735
HOSPITAL	0.771836	0.8119619	0.7873684	0.8055196
DURATION	0.75	0.75	0.6428571	0.75
AGE	0.875	0.8484849	0.877551	0.8598131
ORGANIZATION	0.4370861	0.4697986	0.4129032	0.427673
SET	0	0	0	0
LOCATION-OTHER	0	0	0	0
PHONE	0	0	0	0
Micro-avg.F	0.8658055	0.8906692	0.8961812	0.8868573
Macro-avg.F	0.6783662	0.7088497	0.7049651	0.6947666
Coding Type	**Epochs = 5**	**Epochs = 10**	**Epochs = 15**	**Epochs = 30**
DATE	0.7906977	0.7921502	0.7776616	0.7906977
TIME	0.1517451	0.4004657	0.3109541	0.1613833
DURATION	0.7619048	0.7619048	0.7619048	0.7619048
SET	0	0	0	0
Micro-avg	0.712096	0.7315078	0.7068599	0.7103586
Macro-avg	0.4317993	0.489716	0.4639286	0.43271

4.4 Comparison of Different Model Sizes

Table 6 presents the results of training on the same dataset for ten epochs using models of different sizes. When compared to Table 4, it is evident that the scores slightly improved with thirty epochs of training compared to ten epochs. However, there was no significant increase. This suggests that increasing the number of model sizes can enhance the accuracy of model predictions to some extent.

Table 6. Scores Comparison for 70 m, 410 m, and 1.4 b Models (10 epochs).

Coding Type	70m	410m	1b	1.4b
MEDICALRECORD	0.8632075	0.8777319	0.8745562	0.8787700
PATIENT	0.7987261	0.8859935	0.9456825	0.9459274
IDNUM	0.8512008	0.9510819	0.9430588	0.9579594
DATE	0.9090162	0.9334136	0.9427115	0.9293210
DOCTOR	0.6808652	0.8571428	0.8232737	0.9109430
STREET	0.875513	0.8543956	0.9555237	0.9462990
CITY	0.9118457	0.9338844	0.9379311	0.9475138
STATE	0.889881	0.8494453	0.9759036	0.9697885
ZIP	0.9353932	0.9572650	0.9346591	0.9659091
TIME	0.3202756	0.6966825	0.4459459	0.7487431
DEPARTMENT	0.6919059	0.8119326	0.8160622	0.7878788
HOSPITAL	0.6863742	0.7982222	0.8287342	0.8482587
DURATION	0.75	0.75	0.72	0.72
AGE	0.8349514	0.8800000	0.8823529	0.8761905
ORGANIZATION	0	0.4071856	0.4938272	0.4645161
SET	0	0	0	0
LOCATION-OTHER	0	0	0	0
PHONE	0	0	0	0
Micro-avg.F	0.7843592	0.8829505	0.8756325	0.9084879
Macro-avg.F	0.6137688	0.7000789	0.6976507	0.7186083
Coding Type	**Epochs = 5**	**Epochs = 10**	**Epochs = 15**	**Epochs = 30**
DATE	0.6860805	0.7920585	0.7957215	0.7933403
TIME	0.04605263	0.3175853	0.1227545	0.3463542
DURATION	0.7619048	0.7619048	0.7619048	0.7619048
SET	0	0	0	0
Micro-avg	0.6211579	0.7263592	0.7125595	0.7312019
Macro-avg	0.3798644	0.4707186	0.424776	0.4782775

4.5   Comparing the Impact of Regular Expressions and Post-Processing

From Table 7, we can observe the impact of using regular expressions on the scores. According to this table, using regular expressions is beneficial for score improvement. For example, in the case of 'age,' the score increased from the original 0.6511628 to 0.7022901, and for 'duration,' the score increased from 0.4 to 0.6666667.

Our post-processing involves retaining only one instance of repeatedly predicted data and removing the rest. From this table, it can be inferred that our post-processing is also beneficial for the results.

Table 7. Scores Comparison for 410m Model with and without Post-Processing.

Coding Type	Without Regular Expressions and Post-Processing	Regular Expressions for Age, Duration and without Post-Processing	Regular Expressions for Age, Duration and Post-Processing
MEDICALRECORD	0.87877	0.87877	0.87877
PATIENT	0.9646569	0.9646569	0.9646569
IDNUM	0.952178	0.952178	0.9544376
DATE	0.9028482	0.9028482	0.9287577
DOCTOR	0.8857321	0.8857321	0.886424
STREET	0.9034853	0.9034853	0.9034853
CITY	0.9381018	0.9381018	0.9381018
STATE	0.958209	0.958209	0.958209
ZIP	0.945869	0.945869	0.945869
TIME	0.4060284	0.4060284	0.4877006
DEPARTMENT	0.7701862	0.7701862	0.7898735
HOSPITAL	0.7699918	0.7699918	0.8055196
DURATION	0.4	0.6666667	0.75
AGE	0.6511628	0.7022901	0.8598131
ORGANIZATION	0.3626373	0.3626373	0.427673
SET	0	0	0
LOCATION-OTHER	0	0	0
PHONE	0	0	0
Micro-avg.F	0.8688307	0.8687732	0.8868573
Macro-avg.F	0.5209298	0.5251118	0.6947666
Coding Type	**Epochs = 5**	**Epochs = 10**	**Epochs = 15**
DATE	0.7902787	0.7902787	0.7906977
TIME	0.1611511	0.1611511	0.1613833
DURATION	0.2666667	0.7619048	0.7619048
SET	0	0	0
Micro-avg	0.7084548	0.7098653	0.7103586
Macro-avg	0.3272916	0.4324955	0.43271

4.6 Comparison of Model Training, Prediction Time, and Scores

Epochs Comparison. For the 70m model, the results show little difference among epochs 5, 10, and 15. Contrary to this, the 410m model performs better at epoch 15 than at epoch 10. However, the score difference is minimal and is unlikely to significantly impact the results. For the 1b model, it performs better at epoch 5 than at epochs 10 and 15.

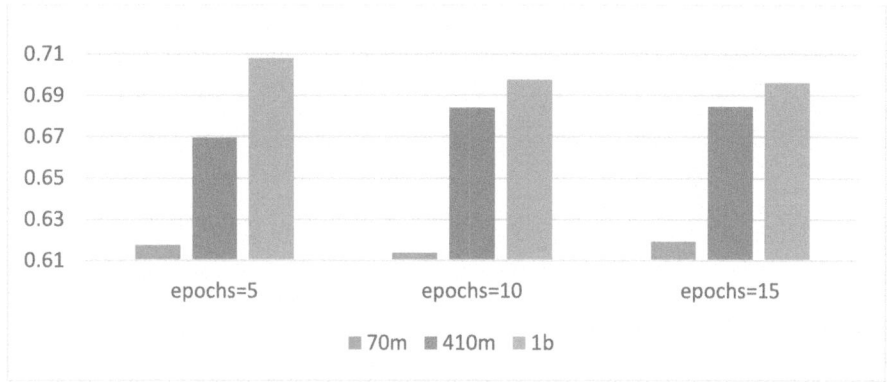

Fig. 1. F-measure comparison for models with different sizes and training epochs.

Model Size Comparison. When comparing the models at epochs 10 and 15, both the 410m and 1b models significantly outperform the 70m model. While the score difference between the 410m and 1b models is small, the 410m model's training and prediction times are notably faster than those of the 1b model (see Fig. 1). Consequently, taking these factors into account, we chose the 410m model. Tables 8, 9 and 10 further compare the required training time, prediction time and the performance comparison for the three models in different model scales.

Table 8. Time for Training the Model.

Model	Epochs = 5	Epochs = 10	Epochs = 15
70m	0:24:27	0:48:33	1:12:38
410m	1:37:44	3:11:00	4:49:40
1b	2:16:11	4:36:55	7:10:46

Table 9. Time for Predicting Results.

Model	Epochs = 5	Epochs = 10	Epochs = 15
70m	0:11:44	0:08:00	0:11:17
410m	7:53:40	5:03:23	5:00:46
1b	9:07:07	7:16:34	11:07:07

Table 10. Score Comparison.

Model	Epochs = 5	Epochs = 10	Epochs = 15
70m	0.617429	0.6137688	0.6191041
410m	0.6692813	0.6838962	0.6844825
1b	0.707866	0.6976507	0.6962177

5 Discussion

This study demonstrated the application of the EleutherAI/pythia-410m-deduped model for enhancing the de-identification of medical information, an essential task in maintaining patient privacy while utilizing EMRs for research. Our findings highlight the model's effectiveness, as evidenced by its performance in the AI CUP 2023 competition. However, the nuanced comparison with previous works, particularly in employing large language models for medical data processing, uncovers a rich landscape of opportunities and challenges.

Comparison with Existing Literature: The use of GPT-3 [2] for summarizing medical literature, as explored in previous studies, shares similarities with our approach, emphasizing the potential of large language models in the medical domain. However, our work distinguishes itself by focusing on de-identification, a critical step towards safeguarding patient information. The comparison reveals that while the core technology can be similar, the application and optimization strategies differ, underscoring the adaptability of large language models to various tasks in healthcare information processing.

Research Findings and Implications: The effectiveness of our method, leveraging a hybrid approach that combines rule-based systems with transformer models, implies a significant step forward in automating the de-identification process. This automation could drastically reduce manual labor and errors, thereby enhancing the efficiency and reliability of privacy protection in EMRs. Moreover, our work contributes to the broader field of clinical NLP, offering insights into handling sensitive information within healthcare data.

Limitations and Future Directions: Despite the promising results, our study is not without limitations. The reliance on a singular dataset, the "OpenDeID," might limit the generalizability of our findings across different data formats and institutions. Additionally, the challenges associated with multi-document summarization observed in GPT-3 suggest potential limitations in our model's ability to process information across multiple

EMRs accurately. Future research could explore the integration of more diverse datasets, the development of more sophisticated models capable of understanding the complex context of medical records, and the exploration of models' scalability and adaptability to different languages and healthcare systems.

Concluding Remarks: The exploration of Pythia for medical information de-identification within the AI CUP 2023 has not only showcased the potential of large language models in enhancing privacy protection but also opened avenues for future research in the domain of clinical NLP. The journey ahead involves addressing the limitations identified, expanding the scope of research, and continuously refining the models to better serve the evolving needs of healthcare information privacy.

This section aims to enrich the paper by providing an in-depth analysis of the research findings, drawing meaningful comparisons with existing literature, and outlining the future trajectory of research in medical information de-identification.

6 Conclusion

Under the 410 m and 1b models, we observed that the score difference is not significant. It's crucial to pay attention to adjusting the training frequency for each model to avoid overfitting issues. When employing models of different sizes, it's essential not only to consider their accuracy but also to weigh their time costs. Finding a relatively better model and adjusting hyper-parameters require careful consideration and trade-offs.

In the end, we utilized the 410 m model with epochs = 10, incorporating LoRA and regular expression to complete this competition. The resulting score is 0.684, which is not exceptionally high. However, for future improvements, we plan to leverage ChatGPT to generate more examples for training, incorporate additional regular expressions corresponding to different categories, or invest more time (without using LoRA) to enhance the accuracy of our model.

Acknowledgments. The authors wish to express their deep appreciation to the SREDH Consortium, the Department of Electronic Engineering at National Kaohsiung University of Science and Technology, and the Department of Biomedical Informatics and Medical Engineering at Asia University, for their organization of the successful competition. Special thanks are reserved for Professor Hong-Jie Dai and his team for their exceptional support and mentorship, which were invaluable throughout the competition.

Disclosure of Interests.   The authors have no competing interests.

References

1. Jonnagaddala, J., Chen, A., Batongbacal, S., Nekkantti, C.: The OpenDeID corpus for patient de-identification. Sci. Rep. **11**(1), 19973 (2021)
2. Shaib, C., Li, M.L., Joseph, S., Marshall, I.J., Li, J.J., Wallace, B.C.: Summarizing, simplifying, and synthesizing medical evidence using GPT-3 (with varying success). arXiv preprint arXiv:230506299 (2023)

3. Liu, J., Gupta, S., Chen, A., Wang, C.-K., Mishra, P., Dai, H.-J., et al.: OpenDeID pipeline for unstructured electronic health record text notes based on rules and transformers: de-identification algorithm development and validation study. J. Med. Internet Res. **25**, e48145 (2023)

4. Moore, C., Bulgarelli, L., Pollard, T., Johnson, A.: Transformer-DeID: de-identification of free-text clinical notes with transformers

5. Chen, A., Jonnagaddala, J., Nekkantti, C., Liaw, S.-T.: Generation of surrogates for de-identification of electronic health records. In: MEDINFO 2019: Health and Wellbeing e-Networks for All: IOS Press, p. 70–3 (2019)

6. Mir, T.H., Yang, H.-P., Chou, Y.-Y., Teng, Y.-C., Liao, W.-H., Lin, Y.-C., et al.: De-identification and temporal normalization of electronic health record notes using large language models: the SREDH/AI-Cup 2023 De-identification Competition. In: Proceedings of the 2024 International Workshop on De-identification of Electronic Medical Record Notes; Kaohsiung, Taiwan: Springer Nature (2024). https://link.springer.com/book/9789819779659

7. SREDH 2024. https://www.sredhconsortium.org/

8. 2024 International Workshop on De-identification of Electronic Medical Record Notes (IW-DMRN) (2024). https://www.sredhconsortium.org/sredh-competitions/sredhai-cup-2023/2024-iw-dmrn

9. Wang, C.-K., Wang, F.-D., Lee, Y.-Q., Chen, P.-T., Wang, B.-H., Su, C.-H., et al.: Principle-based approach for the de-identification of code-mixed electronic health records. IEEE Access. **10**, 22875–22885 (2022)

10. Lee, Y.-Q., Chen, C.-T., Chen, C.-C., Lee, C.-H., Chen, P., Wu, C.-S., et al.: Unlocking the secrets behind advanced artificial intelligence language models in de-identifying Chinese-English mixed clinical text: development and validation study. J. Med. Internet Res. **26**, e48443 (2024)

11. Question answering system for healthcare information based on BERT and GPT. In: Jeong, S.W., Kim, C.G., Whangbo, T.K., (eds.) 2023 Joint International Conference on Digital Arts, Media and Technology with ECTI Northern Section Conference on Electrical, Electronics, Computer and Telecommunications Engineering (ECTI DAMT & NCON). IEEE (2023)

12. Alla, N.L.V., Aipeng, C., Batongbacal, S., Nekkantti, C., Dai, H.-J., Jonnagaddala, J.: Cohort selection for construction of a clinical natural language processing corpus. Comput. Methods Programs Biomed. **1**, 100024 (2021)

13. Dai, H.-J., Lee, Y.-Q., Nekkantti, C., Jonnagaddala, J.: Family history information extraction with neural attention and an enhanced relation-side scheme: algorithm development and validation. JMIR Med. Inform. **8**(12), e21750 (2020)

14. Ding, N., Qin, Y., Yang, G., Wei, F., Yang, Z., Su, Y., et al.: Parameter-efficient fine-tuning of large-scale pre-trained language models. Nat. Mach. Intell. **5**(3), 220–235 (2023)

15. Pythia: a suite for analyzing large language models across training and scaling. In: Biderman, S., Schoelkopf, H., Anthony, Q.G., Bradley, H., O'Brien, K., Hallahan, E., et al. (eds.) International Conference on Machine Learning. PMLR (2023). https://www.sredhconsortium.org/sredh-workshops/2024-iw-dmrn

16. Jonnagaddala, J., et al.: Mining electronic health records to guide and support clinical decision support systems. In: Moon, J.D., Galea, M.P. (eds.) Improving Health Management through Clinical Decision Support Systems, pp. 252–269. IGI Global, Hershey (2016)

17. ISLAB: Privacy protection and medical data standardization competition: decoding clinical cases and letting data tell stories (2023). https://codalab.lisn.upsaclay.fr/competitions/15425

Utilizing Large Language Models for Privacy Protection and Advancing Medical Digitization

Zhu-Jian Ru[1] , Omkar Panchal[3] , Ching-Tai Chen[2] , Jitendra Jonnagaddala[4,5] ,
Hong-Jie Dai[6] , and Sheng-Chun Hsueh[1(✉)]

[1] High Performance Distributed Systems Lab, Department of Electrical Engineering, National Kaohsiung University of Science and Technology1, No. 415, Jiangong Road, Sanmin, Kaohsiung, Taiwan R.O.C.
{f112154143,f112154136}@nkust.edu.tw
[2] Department of Bioinformatics and Medical Engineering, Asia University, Taichung, Taiwan
ctchen@asia.edu.tw
[3] CGD Health Pvt. Ltd., Hyderabad, Telangana, India
omkar@cgdhealth.com
[4] University of New South Wales, Sydney, Australia
z3339253@unsw.edu.au
[5] NMC Royal Hospital, Khalifa City, Abu Dhabi, United Arab Emirates
[6] Intelligent System Lab, Department of Electrical Engineering, National Kaohsiung University of Science and Technology1, No. 415, Jiangong Road, Sanmin, Kaohsiung, Taiwan R.O.C.
hjdai@nkust.edu.tw

Abstract. This research explores the utilization of artificial intelligence (AI) language generation models for the de-identification of medical case narratives, effectively anonymizing patient identifiers to prevent unauthorized disclosure of sensitive health information. This ensures that patient data remains unrecognizable even when accessed, offering two significant advantages: compliance with legal standards forbidding the indiscriminate release of medical records, and bolstered trust in healthcare providers which encourages patients to share necessary details for therapeutic and investigative purposes. Historically, de-identification processes necessitated the manual examination of records by specialists to identify and sanitize personal identifiers, it is a time-consuming task with potential for error. The digitalization of medical records now offers avenues for automated de-identification, simultaneously enhancing data accuracy and facilitating the ethical utilization of extensive datasets in healthcare research, public health strategy, and policymaking. The emergence of sophisticated generative AI technologies has augmented the capacity for comprehensive de-identification. Platforms such as ChatGPT and Bing have evolved to address complex human inquiries. Yet, employing these tools without proper safeguards potentially compromises de-identification objectives by risking data exposure. To advance the integrity of privacy measures and maintain data governance, this study fine-tunes a pre-established vast language AI model, Pythia, for localized training with state-of-the-art technology with rigorous privacy assurance.

Keywords: Large language models · Sensitive health information · Fine tuning · QLora

© The Author(s), under exclusive license to Springer Nature Singapore Pte Ltd. 2025
J. Jonnagaddala et al. (Eds.): IW-DMRN 2024, CCIS 2148, pp. 177–188, 2025.
https://doi.org/10.1007/978-981-97-7966-6_13

1 Introduction

De-identification primarily involves two core methods: pseudonymization [1], which replaces personal information with a fictitious alias or code, thus concealing the direct display of identity traits. Although pseudonymized data alone does not reveal specific individuals, it may still be traceable to a particular person when combined with supplementary information. This contrasts with anonymization [2], which is utilized in this research, where patient identity information is entirely removed, rendering the data unlikable to any specific individual. It's crucial to establish trust, de-identify sensitive health information (SHI), and promote responsible and ethical use of healthcare data for research analysis. Given how important it is to keep SHI private [3], the AI Cup 2023-Privacy Protection and Medical Data Standardization competition [4] and the 2024 International Workshop on the de-identification of Electronic Medical Record Notes (IW-DMRN) were organized [5, 6]. AI Cup 2023 offers a platform to address issues with data standards and privacy protection in medical records.

Common privacy preservation techniques [7–9] include K-anonymity, which modifies and abstracts personal data to ensure that each individual is indistinguishable from at least K-1 others in certain key attributes, thus safeguarding the privacy of individuals from identification; L-diversity, which extends the principle of K-anonymity by ensuring diversity within the sensitive attributes, guaranteeing that each group with identical key properties contains at least L distinct sensitive attribute values to prevent attackers from inferring personal sensitive information through ancillary data; and T-closeness, which maintains that the distribution of sensitive attributes within groups sharing certain key properties should adhere to specific 'closeness' constraints relative to the overall dataset distribution, complicating the task for attackers attempting to deduce an individual's SHI. Nonetheless, the patient data we handle is often irregular and interspersed with a myriad of noise, and the formatting between case reports and medical records is not standardized, making the aforementioned methods unsuitable. Consequently, AI assistance is required to recognize and label names, locations, hospital names, dates, and birthdates and to normalize dates and times [10, 11].

Current prominent language models like Generative Pre-trained Transformer (GPT)-3, GPT-4, LLaMA (Large Language Model Meta AI), BERT [12], and RoBERTa [13] have required vast amounts of data for training, which is often beyond the reach of smaller companies or developers and requires substantial computational resources for model training. Therefore, we use pre-trained models and fine-tune them [9, 10]. Open-source models available today include LLAMA, BLOOM [16] and Pythia [17]. We have chosen the Pythia model because it is specifically designed to address visual questions [18], including extracting information [19] from text and visual signals from images. Our experimentation has revealed that using models with larger parameters typically yields better results. However, upon deploying the Pythia 1b model, we encountered a performance bottleneck, prompting us to employ LoRA (Low-Rank Adaptation of large language models) [20] for fine-tuning parameters. Despite significant improvements by LoRA in reducing storage requirements, substantial GPUs are still necessary for loading and training models. Therefore, we further applied QLoRA (Quantized LoRA) [21] to compress information, reducing memory utilization on GPUs. In the next section, we delve into the intricacies of the Pythia model's parameters, as well as offer a comparative analysis between LoRA and QLoRA. We then focus on data processing for the model training procedure and report the results in the Results section. The Discussion section is dedicated to the analysis of the experimental results. Finally, we summarize our findings and discusses potential future research directions.

2 Methods

2.1 Pythia Models and LoRA

In the work, we employed large language models (LLMs) based on the Pythia as the base model for fine-tuning. The Pythia framework is a component-based constructed architecture, containing multiple models that are all based on the Transformer architecture [22]. The structure of Pythia is divided into three main parts: the input encoder, the decoder, and the predictor. The input encoder is responsible for converting the input sequence into a series of feature vectors that carry semantic information from the input sequence. Within Pythia, the input encoder uses multiple layers of Transformer encoders, each layer including several self-attention mechanisms and feed-forward neural networks. The decoder then transforms the feature vectors outputted by the encoder into an output sequence, also employing multiple layers of Transformer decoders. In Pythia, the decoder incorporates a masking mechanism to ensure that future information is not referenced when generating the output sequence. The predictor is tasked with converting the feature vectors output by the decoder into the final predicted results, such as words or tokens. In Pythia, the predictor utilizes a fully connected layer and converts the outputs into probability distributions through the softmax function. Such architecture enhances the model's performance and adaptability. Table 1 shows the various models provided in the Pythia framework.

Table 1. Pythia model Comparison.

Model Size	Non-Embedding Params	Layers	Model Dim	Heads	Learning Rate
70M	18,915,328	6	512	8	1.0×10^{-4}
160M	85,056,000	12	768	12	6.0×10^{-4}
410M	302,311,424	24	1024	16	3.0×10^{-4}
1.0B	805,736,448	16	2048	8	3.0×10^{-4}
1.4B	1,208,602,624	24	2048	16	2.0×10^{-4}
2.8B	2,517,652,480	32	2560	32	1.6×10^{-4}
6.9B	6,444,163,072	32	4096	32	1.2×10^{-4}
12B	11,327,027,200	36	5120	40	1.2×10^{-4}

To fine tune the Pythia models, we applied the QLoRA method, which can improve the efficiency and scalability of the self-attention mechanism within the original transformer architecture. The traditional self-attention mechanism requires the calculation of pairwise attention weights among all elements in the input sequence, and the computational complexity increases quadratically with the length of the sequence, which causes efficiency issues when processing long texts or when there is a desire to build larger models. Furthermore, in the traditional self-attention mechanism, it is necessary to calculate the attention weight between every pair of elements in the sequence, leading to a significant increase in computational and storage requirements with sequence length. Compared to the fully fine tuning all of the parameters in the Pythia models, through QLoRA, we introduce a small set of trainable parameters to update the model while keeping other parts of the model fixed. This dramatically reduces computation complexity and improves memory efficiency. In our implementation, we specifically employed a 4-bit quantization strategy and the normal float (NF4) data format to quantize model parameters, optimizing memory usage while maintaining good training performance. By using the QLoRA technique to improve the fine-tuning procedure of the Pythia model, we can not only avoids memory peak issues when processing long sequences in small batches but also support larger model sizes while maintaining predictive performance under the premise (Table 2).

Table 2. Comparison between LoRa and QLora

Feature	LoRa	QLoRa
Trainable Parameters	Small set of trainable parameters	Parameter updates using 4-bit quantization
Parameter Updates	Keeps most of the model parameters fixed	Freezes most of the model parameters; only updates the 4-bit quantized LoRA parameters
Training Precision Requirement	Require 16-bit precision	Use 4-bit precision, specifically NF4 quantization
Quantization Strategy	–	Introduces NF4 quantized data type, saving an average of about 0.37 bits
Memory Management	–	Uses a paging optimizer to manage memory peaks, suitable for long sequences and large batch processing

Our greatest difference lies in our use of QLoRA, which only employs parameter updates through 4-bit quantization. Most of the model parameters are frozen, and only the 4-bit quantized LoRA parameters are fine-tuned. By introducing the NF4 quantized data type, this data type design ensures that all parameters are stored in four bits before any actual computation takes place. During computation, they are converted to BF16 or FP16 as needed. This approach not only drastically reduces memory consumption during model operation but also ensures computational efficiency.

To further illustrate why NF4 was chosen, let us introduce the method of double quantization. This technique involves the re-quantization of quantized constants, usually reducing the originally 32-bit represented quantization constants down to 8-bit integers (INT8) to diminish the extra memory overhead. Although such technology can save memory, it may also increase the time cost of model inference, as it requires restoring parameters during inference in each model layer. Therefore, we chose to use NF4 in QLoRA, considering that NF4 can save memory while maintaining the efficiency of model inference. The design of NF4 ensures that before actual computation, all parameters are stored in four bits and are converted to Brain floating 16 (BF16) or half precision 16-bit floating point precision types (FP16) as needed during computation, ensuring computational efficiency. Based on our experimental results described later, we observed that using NF4 enables us to fine-tune larger Pythia models without hindering the overall performance comparing with fully fine-tuning of the models.

2.2 Data Processing and the Model Training Method

We used the dataset released by the AI Cup 2023 for developing our models. The dataset is based on the OpenDeID corpus provided by SREDH Consortium [19, 20]. The released dataset comprises 3,244 pathology reports from 1,833 distinct cancer patients across four urban hospitals in Australia. This corpus contains 38,414 instances of SHI in total. Two annotators worked together in three different experimental scenarios to generate this corpus [22–24].

Preprocessing: Our training set synthesizes the first batch of training data, the validation set, and the second batch of training data, combining structured text data with corresponding answer annotations. After integration, these training data and their answer annotations are read. First, useful information is extracted from the answer files and converted into dictionary format. Then, each document is read, corresponding to the answer annotations, and they are integrated into a specific format by paragraph.

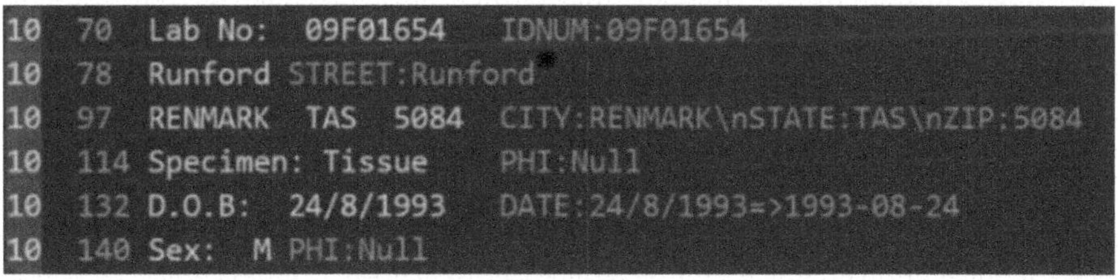

```
10  70  Lab No:  09F01654   IDNUM:09F01654
10  78  Runford STREET:Runford
10  97  RENMARK  TAS  5084  CITY:RENMARK\nSTATE:TAS\nZIP:5084
10  114 Specimen: Tissue    PHI:Null
10  132 D.O.B:  24/8/1993   DATE:24/8/1993=>1993-08-24
10  140 Sex:  M PHI:Null
```

Fig. 1. Identification of SHI

As shown in Fig. 1, the integrated file contains document names, line numbers, original text content, and corresponding annotation information.

For instance, SHI within a document is marked according to its category. If an item's content involves a date or time, it is appended with "= > normalized content" and retains its start and end index values from the original document. These transformed data are organized into a new file for use by the data loader (DataLoader). The role of DataLoader is to make the data more suitable for feature extraction and training in machine learning algorithms. It begins by reading annotated information from answer files and then pairs it with the corresponding document. Processing these data, DataLoader creates a formatted file containing text and annotation information, maintaining consistency, and providing a more comprehensive and accurate dataset for model training.

Post-processing: The read_file function is first used to read text files, followed by the process_valid_data function to format these files, extracting filenames, text boundaries, and content. These data are then written into a new output file. Afterward, a new custom list is defined, including various SHI categories like patient names, doctors' names, and addresses. The get_anno_format function processes the predicted SHI labels, which are based on the model's predictions within a given sentence. The function searches for the original sentence location and establishes corresponding output, including entity type, start and end positions within the text. If these entities are dates or times, the function also performs normalization. Finally, the predict function utilizes the model and tokenizer to generate predictions, organizing the results to identify SHI within the text and creating formatted annotation information.

For the PHONE category, which has less training data, we used a normalization capture. Due to the smaller amount of data, we took the approach of manual checking, with the normalization regular expression being '\d {4} \d {4}'.

2.3 Data Analysis and Experimental Setup

During the training of our model, we started with the Pythia framework, whose flexibility and modularity allowed us to quickly combine different machine learning model components to adapt to a variety of research questions and datasets. Building upon the Pythia framework, we further integrated PyTorch, which provides powerful computational graph capabilities and an automatic differentiation system. The dynamic nature of PyTorch's computing graphs made it possible for us to adjust the computation workflow more flexibly during training. To further increase computational efficiency, we moderately adopted mixed-precision training, transferring some computations from 32-bit to 16-bit floating-point numbers. This allowed us to speed up training and reduce memory usage while ensuring model convergence and final accuracy through stable gradient scaling methods.

Finally, by combining the LoRA technique and extending it based on Pythia and PyTorch, we introduced QLoRA technology into PyTorch, significantly improving the computational efficiency and scalability of the self-attention mechanism within the transformer architecture. Through the QLoRA approach, which involves updating and adjusting a small subset of the parameters within the network through 4-bit quantization while keeping most other model parameters fixed, we effectively balanced training precision with model memory usage. Using the NF4 quantization strategy, we were able to significantly reduce the required memory footprint without affecting inference performance, allowing the model to remain efficient even when processing long sequence data or handling large batch sizes. We also performed special optimizations on key components within the model. In the self-attention mechanism's query, key and value parts, we reduced memory usage through quantization while ensuring precise calculations. In the dense fully connected layers, we enhanced the learning process to achieve limited parameter updates while capturing complex features. For the dense_h_to_4h hidden layer expansion operation, we used LoRA to ensure memory efficiency and the necessary computations co-exist. Dense_4h_to_h achieved layer reduction expansion and maintained information transmission efficiency through quantization.

Throughout the entire training process on top of NF4, we reduced memory requirements, accommodating the needs for long sequences and large batch processing. Particularly during forward propagation computations, we precisely quantized and updated important parameters such as query_key_value, dense, dense_h_to_4h, and dense_4h_to_h in 4-bit precision, further reducing computational difficulty and memory demands. We continuously monitored memory usage to ensure the model operated stably under high-load challenges.

In our testing environment, which includes Ubuntu 20.04.6 LTS as the operating system, Python 3.9.17 as the programming language, and an RTX 3090ti GPU, we employed models Pythia 70m, Pythia 170m, and Pythia 1b. However, the Pythia 1b model sometimes faced out of memory issues due to inadequate hardware capabilities. To address this, we opted for using LoRA for fine-tuning the 1 b model.

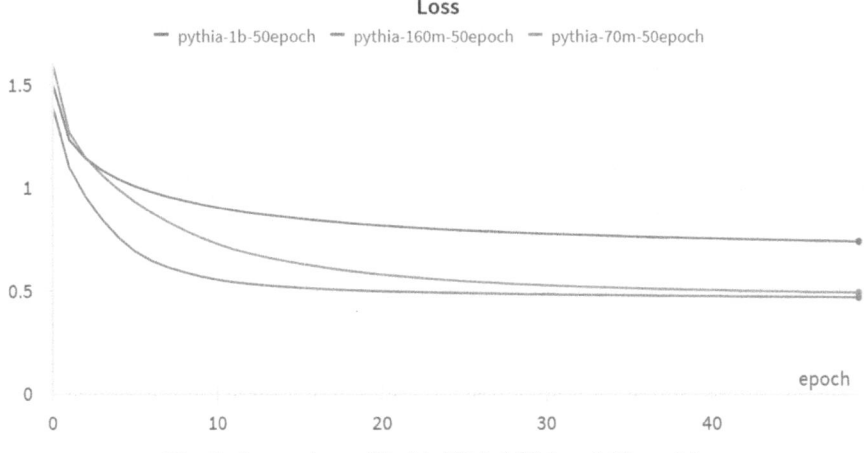

Fig. 2. Loss values of Pythia 70M, 160M, and 1B models

Figure 2 shows the loss values generated by Pythia models of 70M, 160M, and 1B with QLora. In our experiments, we tested over 50 epochs. From Fig. 2, we can observe that the model with 70M parameters begins to show a gradually flattening curve starting from the 20th epoch, indicating that the model is starting to converge. Meanwhile, the 160M model starts to converge around the 15th epoch, gradually approaching a loss value of 0.5. If we follow the trend, it can be inferred that models with a larger number of parameters tend to converge faster and exhibit a greater reduction in loss. However, the 1B model, after employing QLora, shows a slower convergence rate compared to full-parameter training and reaches a minimum loss of approximately 0.8, which is higher than the other two models.

3 Results

Table 3 shows our results for the AI CUP Privacy Protection and Standardization of Electronic Medical Record Competition, where the Pythia 1B model outperforms the other two models in terms of f-measure, with a score of 0.758. In comparison, the 70M model scored 0.67, and the 160M model scored 0.70. As shown in Table 5 in the prediction results, the "DOCTOR" and "STREET" categories, despite higher precision scores, had lower recall scores. This indicates that the model missed many actual positives in these two categories. For the TIME category, it was observed that the content captured only included the date and lacked the time. From the prediction examples shown in Fig. 4. It is known that TIME did not capture the content correctly in Task One, resulting in no score obtained in the normalization task (Fig. 3 and Table 4).

Table 3. The results of the Pythia models of three varied sizes (70M, 160M, and 1B)

Model	Coding Type	Precision	Recall	F-measure	Support
Pythia 70m	Macro-avg	0.7941052	0.5898103	0.6768786	13007
Pythia 160m	Macro-avg	0.7710106	0.6523982	0.7067624	13007
Pythia 1B-Qlora	Macro-avg	0.8371984	0.6939132	0.7588513	13007

Table 4. Task one prediction Score of Pythia 1 b with lora

Coding Type	Precision	Recall	F-measure	Support
DATE	0.8151742	0.751525	0.7820567	2459
DURATION	0.5	0.0833333	0.1425871	12
TIME	0	0	0	470
SET	1	0.4	0.57144286	5
Micro-avg.	0.8114862	0.6283095	0.7082456	2946
Macro-avg.	0.5787935	0.3087146	0.40266	2946

Table 5. Task two prediction score of Pythia 1 b with lora

Coding Type	Precision	Recall	F-measure	Support
MEDICALRECORD	0.7916667	0.9156626	0.89162	747
PATIENT	0.88251	0.9231843	0.9023891	716
IDNUM	0.9669421	0.8278302	0.8919949	2120
DATE	0.952121	0.9219195	0.9367769	2459

(*continued*)

Table 5. (*continued*)

Coding Type	Precision	Recall	F-measure	Support
DOCTOR	0.9372043	0.6549444	0.7710545	3327
STREET	0.9266409	0.6976744	0.79602	344
CITY	0.9637883	09276139	0.9453552	373
STATE	0.993921	0.9849398	0.98941	332
ZIP	0.994302	0.9886686	0.9914773	353
DEPARTMENT	0.8593351	0.8019093	0.8296297	419
HOSPITAL	0.8558952	0.8180301	0.8365344	1198
TIME	0.131579	0.0212766	0.0366300	470
AGE	0.8571429	0.7058824	0.7741936	51
ORGANIZATION	0.4565217	0.5675676	0.5060241	74
SET	1	0.4	0.5714286	5
LOCATION-OTHER	1	0.1666667	0.257143	6
PHONE	1	1	1	1
Micro-avg.	0.9134632	0.785577	0.8447071	13007
Macro-avg.	0.8371984	0.6939132	0.7588513	13007

Fig. 3. Missing TIME results

4 Discussion

In this research, we conducted a comparative analysis of three distinct models. Initially, the 70M model demonstrated promising capabilities in the initial stages of testing; however, it faced challenges in maintaining its performance as the complexity of tasks increased. This trend underscores the notion that while smaller-scale models can excel in specific applications, their efficacy tends to diminish in more complex scenarios. In contrast, the 160M model exhibited a commendable balance between efficiency and effectiveness across a broader array of tasks. Its increased parameter size, compared to the 70M model, aided in better managing complex requirements, particularly in the areas of privacy protection and standardization of electronic medical records. The 1B model, despite its higher accuracy, required significantly more time and computational resources in local testing. Its vast expansion in parameter size enabled it to outperform the other models in most evaluation tasks. Its ability to process and assimilate large datasets was especially effective in identifying and protecting sensitive information within EMRs, as well as ensuring their standardization, highlighting its superior capability in handling high-stakes data management challenges.

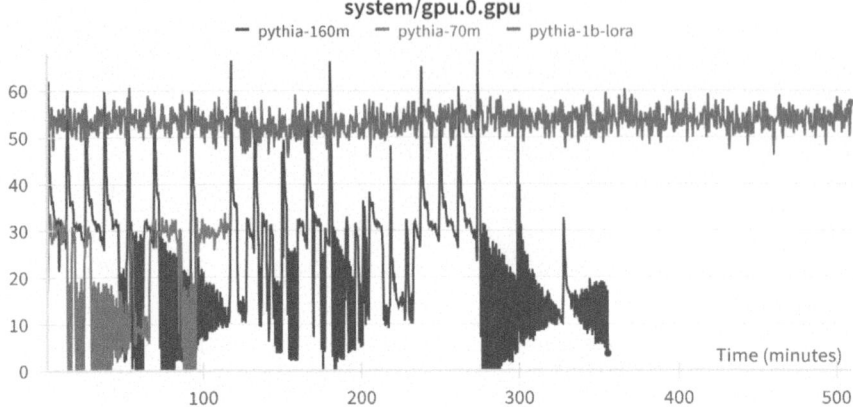

Fig.4. GPU Resources and utilized time by three models.

Figure 4 shows the time spent and GPU resources consumed by the three models. The Pythia 1b with lora model had 4,194,304 parameters trained during the training stage, taking 8 h and 29 min, while the 70m and 170m models respectively took 1 h and 55 min, and 5 h and 55 min for training.

Throughout the training and testing phases, we observed that, in terms of loss metrics, the Pythia 1B model initially lagged its smaller counterparts, the 70M and 160M models. Despite this early underperformance in loss, the ultimate evaluation revealed a different story: the 1B model's effectiveness outstripped that of both the 70M and 160M models. This phenomenon can be attributed to the superior capability of larger models to encapsulate and recall intricate data patterns. It is crucial to note that the evaluation of a model's final performance often relies on metrics that do not necessarily align with the loss function utilized during its training phase. Thus, despite the 1B model exhibiting higher loss values initially, its superior performance on key evaluation metrics relevant to the task such as accuracy and F1 score underscores its effectiveness.

In this research, we focus on testing the depth of model parameters. Through this approach, we aim to reveal how the choice of parameters significantly affects the performance and efficiency of the model. By delving into various parameter settings, we can better understand how the model performs on specific tasks and its capability in handling complex data. In future work, we could expand our research scope to investigate how different network architectures and hierarchies affect model performance. This includes exploring networks that are either deeper or shallower and examining how models' understanding of data can be improved through enhancement and preprocessing techniques, such as data augmentation in preprocessing.

5   Conclusion

Three models were compared in this study: the 70M, 160M, and 1B, and the results showed how well each performed on different tasks. The 160M model achieved a compromise between efficacy and efficiency, whereas the 70M model showed early promise but suffered with complexity. Even with its higher resource requirements, the 1B model fared better than the others, especially when handling high-stakes data issues. The 1B model performed poorly at first, but its improved recall of complex data patterns demonstrated its efficacy. Our research emphasizes how important parameter selection is to model performance and recommends further investigation into network topologies and preprocessing methods in the future.

Acknowledgments. We would like to express our gratitude to the organizers of the AI CUP 2023 Autumn competition (https://codalab.lisn.upsaclay.fr/competitions/15425), Our appreciation also goes to the University of New South Wales, Lowy Cancer Research Centre and Health Science Alliance Biobank for granting access to the dataset utilized in this study. Furthermore, we acknowledge the SREDH Consortium (https://www.sredhconsortium.org/) and the IW-DMRN workshop (https://www.sredhconsortium.org/sredh-competitions/sredhai-cup-2023/2024-iw-dmrn) for their invaluable contributions to our research activities. to our research endeavors and the SREDH Consortium (https://www.sredhconsortium.org/) Translational Cancer Bioinformatics working group in accessing the OpenDeID corpus v2 Dataset. JJ was funded by the Australian National Health and Medical Research Council (GNT1192469) and supported by 2022 Google's research grants and cloud computing resources (GCP19980904), as well as the Research Technology Services at University of New South Wales Sydney and NVIDIA's academic hardware grant programs.

Disclosure of Interests. The authors have no competing interests to declare that are relevant to the content of this article.

References

1. Volodina, E., Dobnik, S., Tiedemann, T.L.M, Vu, X.-S.: Grandma Karl is 27 years old – research agenda for pseudonymization of research data. In: 2023 IEEE Ninth International Conference on Big Data Computing Service and Applications (BigDataService). IEEE, Athens, Greece, pp. 229–233 (2023)
2. Pawar, A., Ahirrao, S., Churi, P.P.: Anonymization techniques for protecting privacy: a survey. In: 2018 IEEE Punecon. IEEE, Pune, India, pp. 1–6 (2018)
3. Liu, H.: Research on privacy protection framework design and key technologies in large data environment. In: 2019 International Conference on Robots & Intelligent System (ICRIS). IEEE, Haikou, China, pp. 327–330 (2019)
4. Du, J.: Research on enterprise information security and privacy protection in big data environment. In: 2021 3rd International Conference on Machine Learning, Big Data and Business Intelligence (MLBDBI). IEEE, Taiyuan, China, pp. 324–327 (2021)
5. Wang, C.: Research on the protection of personal privacy of tourism consumers in the era of big data. In: 2018 International Symposium on Computer, Consumer and Control (IS3C). IEEE, Taichung, Taiwan, pp. 428–431 (2018)
6. Warusawithana, S.P., Perera, N.N., Weerasinghe, R.L., Hindakaraldeniya, T.M., Ganegoda, G.U.: Layout aware resume parsing using NLP and rule-based techniques. In: 2023 8th International Conference on Information Technology Research (ICITR). IEEE, Colombo, Sri Lanka, pp. 1–5 (2023)

7. Kavita, M., Singh, H.: Utilizing mixture methods for classifier in nlp: an essential considera-tion. In: 2023 International Conference on Artificial Intelligence and Smart Communication (AISC). IEEE, Greater Noida, India, pp. 422–426 (2023)

8. Gangavarapu, A.: LLMs: a promising new tool for improving healthcare in low-resource nations. In: 2023 IEEE Global Humanitarian Technology Conference (GHTC). IEEE, Radnor, PA, USA, pp. 252–255 (2023)

9. Srar, J.A., Chung, K.-S., Mansour, A.: LLMS adaptive beamforming algorithm implemented with finite precision. In: 2012 20th Telecommunications Forum (TELFOR). IEEE, Belgrade, Serbia, pp. 303–306 (2012)

10. Ma, L., et al.: LLMs with user-defined prompts as generic data operators for reliable data processing. In: 2023 IEEE International Conference on Big Data (BigData). IEEE, Sorrento, Italy, pp. 3144–3148 (2023)

11. Biderman, S., Schoelkopf, H., Anthony, Q., et al.: Pythia: a suite for analyzing large language models across training and scaling (2023). https://doi.org/10.48550/ARXIV.2304.01373

12. Li, B., Li, W., Zhao, D.: Multi-scale feature based medical image classification. In: Proceed-ings of 2013 3rd International Conference on Computer Science and Network Technology. IEEE, Dalian, China, pp. 1182–1186 (2013)

13. Bhavya, B., et al.: Exploring large language models for low-resource IT information extrac-tion. In: 2023 IEEE International Conference on Data Mining Workshops (ICDMW). IEEE, Shanghai, China, pp. 1203–1212 (2023)

14. Hu, E.J., et al.: LoRA: Low-rank adaptation of large language models (2021)

15. Dettmers, T., Pagnoni, A., Holtzman, A., Zettlemoyer, L.: QLoRA: efficient finetuning of quantized LLMs (2023)

16. Mir, T.H., Yang, H.P., Chou, Y.Y., Teng, Y.C., Liao, W.H., Lin, Y.C., et al.: De-identification and temporal normalisation of the electronic health record notes using large language models: the SREDH/AI-Cup 2023 de-identification competition. In: 2024 International workshop on de-identification of electronic medical record notes. Kaohsiung, Taiwan: Springer Nature (2024)

17. ISLAB. Privacy protection and medical data standardization competition: decoding clinical cases and letting data tell stories (2023). https://codalab.lisn.upsaclay.fr/competitions/15425

18. Consortium, S.: 2024 International workshop on de-identification of electronic medical record notes (IW-DMRN) (2024). https://www.sredhconsortium.org/sredh-competitions/sre dhai-cup-2023/2024-iw-dmrn

19. Consortium, S.: Secure research environment for digital health (SREDH) consortium. https://www.sredhconsortium.org/

20. Jonnagaddala, J., Chen, A., Batongbacal, S., Nekkantti, C.: The OpenDeID corpus for patient de-identification. Sci. Rep. 11(1), 19973 (2021)

21. Chen, A., Jonnagaddala, J., Nekkantti, C., Liaw, S.T.: Generation of surrogates for de-identification of electronic health records. In: MEDINFO 2019: Health and Wellbeing e-Networks for All, pp. 70–73. IOS Press (2019)

22. Alla, N.L.V., Chen, A., Batongbacal, S., Nekkantti, C., Dai, H.J., Jonnagaddala, J.: Cohort selection for construction of a clinical natural language processing corpus. Comput. Methods Programs Biomed. Update 1, 100024 (2021)

23. Liu, J., et al.: OpenDeID pipeline for unstructured electronic health record text notes based on rules and transformers: de-identification algorithm development and validation study. J. Med. Internet Res. 25, e48145 (2023)

24. Jonnagaddala, J., et al.: Mining electronic health records to guide and support clinical decision support systems. In: Moon, J.D., Galea, M.P. (eds.) Improving Health Management through Clinical Decision Support Systems, pp. 252–269. IGI Global, Hershey (2016)

Comprehensive Evaluation of Pythia Model Efficiency in De-identification and Normalization for Enhanced Medical Data Management

Yen-Cheng Cho[1]([✉]) [iD], Yu-Jie Yang[2] [iD], Yu-De Liu[1] [iD], Tung-Sheng Tsao[3] [iD], and Min-Jain Lee[4] [iD]

[1] Applied Intelligence Lab, Department of Electrical Engineering, National Kaohsiung University of Science and Technology, Kaohsiung 807618, Taiwan
{F112154137,F112154141}@nkust.edu.tw

[2] Electromagnetic Sensing Control and AI Computing System Laboratory, Department of Electrical Engineering, National Kaohsiung University of Science and Technology, Kaohsiung 807618, Taiwan
F111154166@nkust.edu.tw

[3] Multimedia Protection and Interaction Laboratory, Department of Electrical Engineering, National Kaohsiung University of Science and Technology, Kaohsiung 807618, Taiwan
F112154171@nkust.edu.tw

[4] GPS/ Image/ Satellite Applications Lab, Department of Electrical Engineering, National Kaohsiung University of Science and Technology, Kaohsiung 807618, Taiwan
F112154169@nkust.edu.tw

Abstract. This study presents our work developed for the AICUP2023-privacy protection and standardization of electronic medical record text notes. Our work focuses on exploring the efficiency of the Pythia model developed by the EleutherAI community applying it to the de-identification and normalization problems. The core objective of the research outcome is to achieve a dual purpose: on one hand, to protect patient privacy, and on the other hand, to enhance the automation and efficiency of medical data management, while striving to reduce manual de-identification errors caused by human factors. To this end, we will not only examine the standard configuration of the Pythia model but also delve into its performance under different parameter settings, to comprehensively evaluate and compare its effectiveness and results in handling sensitive medical data. Our experiment results highlight that merely increasing the model size might not be sufficient to yield the expected benefit for recognizing protected health information from text. However, for the task of temporal information normalization, the large language model demonstrates a significant performance improvement after fine-tuning.

Keywords: Pythia Model · protect patient privacy · De-Identification

J. Jonnagaddala et al. (Eds.): IW-DMRN 2024, CCIS 2148, pp. 189–201, 2025.
https://doi.org/10.1007/978-981-97-7966-6_14

1 Introduction

With the rapid advancement of artificial intelligence and natural language processing technology, large pre-trained models like ChatGPT have shown significant potential across various domains. From security and privacy protection to medical information extraction and evaluating the intelligence levels of language models, recent research has unveiled innovative applications and challenges of these technologies. Wu X and colleagues investigated the security and privacy issues related to ChatGPT [1], while Hu D and team demonstrated its capabilities in few-shot learning [2] Wu T and others provided an overview of ChatGPT's development [3], and Qu Y explored text generation systems based on new corpora [4]. Singh SK and colleagues compared ChatGPT with Google Bard AI [5], with studies focusing on de-identification of electronic health records [6] and psychometric analysis of the general intelligence factor in language models [7], offering crucial insights into understanding and applying these advanced models.

In this study, we presented our work developed for the AI-Cup 2023 de-identification competition [8–11, 14]. Using pre-trained models, we employed natural language processing (NLP) technologies to precisely extract sensitive health information (SHI) from electronic medical records (EMR) text notes [13]. This approach allowed us to effectively anonymize such information, safeguarding individual privacy and mitigating the risk of data breaches. Moreover, we employed the pre-trained model developed by the EleutherAI community, named Pythia. Despite this model's wide parameter range from 70M to 12B, our research specifically utilized the Pythia-1b-dedupe model as our primary experimental tool. Predominantly based on the GPT-NeoX model and underpinned by the Transformer model architecture, Pythia presents an extensive array of tunable parameters, furnishing a robust framework for our experimental endeavors.

2 Methods

2.1 Pre-processing Methods

We participated in the development of AICUP2023-privacy protection and standardization of EMR text notes [10] and obtained relevant data, as shown in Fig. 1 from the SREDH platform [8, 9]. We trained our model using the OpenDeID v2 corpus consists of 3,244 reports out of which 2100 reports are from OpenDeID v1 corpus proposed by Jitendra Jonnagaddala et al. [12]. This corpus contains anonymized real medical record data, which we used to train and evaluate our model, aiming to enhance the protection of patient privacy [1].

```
Episode No:  09F016547J IDNUM: 09F016547J
091016.NMT       MEDICALRECORD: 091016.NMT
SIZAR, HOWARD   PATIENT: SIZAR, HOWARD
Lab No:  09F01654       IDNUM: 09F01654
Runford STREET: Runford
RENMARK  TAS  5084       CITY: RENMARK\nSTATE: TAS\nZIP: 5084
Specimen: Tissue         PHI: NULL
D.O.B:  24/8/1993        DATE: 24/8/1993=>1993-08-24
Sex:  M PHI: NULL
Collected: 28/08/2013 at 08:26  TIME: 28/08/2013 at 08:26=>2013-08-28T08:26
```

Fig. 1. Pre-processed training dataset.

2.2   Model Infrastructure and Training Procedure

The Transformer is a type of deep learning model, suitable for applications in language models (LMs). The core concept of the Transformer model is the attention mechanism, designed to process all information in a sequence. It incorporates a multi-head self-attention mechanism in the model, enhancing its ability to capture long-distance dependencies. The Transformer architecture has been proven to enable a better understanding of the context within the text, as it considers not just the order of words but also the significance of each word about others. This capability allows it to perform well in tasks involving natural language contexts and texts.

In our training procedure, we used the Pythia-1b-dedupe model, which accepts a sentence as the input, pays attention to key phrases, and generates its predictions. The generated results include the texts of SHIs mentioned in the given sentence, the types of the SHIs and the normalized values if the SHI is related to date. We opted to use the Pythia-1B-deduped model released by the EleutherAI community for the competition because based on our experiments, we found that the Pythia-1B-deduped model outperformed other Pythia models with a smaller number of parameters. Although the performance of the 1b model is closely followed by Pythia-410m-dedup, the latter's prediction accuracy in DURATION was not as expected, leading us to choose Pythia-1B-deduped for this competition. Further details will be provided in the Results section. Furthermore, the choice of the Pythia-1B-deduped model was due to its high degree of stability among many pre-trained models we tried, especially in certain key aspects where it exhibited superior performance. For example, it tends to have fewer anomalies in prediction results compared to other models, helping us avoid unpredictable issues during the competition and ensuring smooth progress. Figure 1 shows the preprocessed dataset that is used to fine-tune the Pythia model, which includes the paragraph contents, and the corresponding annotated SHIs.

In our preliminary experiment with the use of Pythia-410m-dedupe model, we examined the dataset and manually improved the annotations of the released corpus by removing or correcting annotation errors observed in the dataset. Furthermore, to optimize the model's predictive performance and ensure focus on key information, preprocessing was conducted during the training process. The main aim of this task was to reduce the model's exposure to unnecessary information during prediction. To this end, parts of the original data were trimmed to cleanse the training data and enhance the model's precision.

We then switch to the Pythia-1B-dedupe version for training. This approach allowed us to focus on processing and optimizing prediction results, thereby achieving the goal of enhancing overall performance. The key lies in leveraging the predictive capabilities of the Pythia-1B-dedupe model, while eliminating potential errors through data cleaning and correction, thereby further enhancing the model's performance and reliability.

2.3   Post-processing Method

The goal of the post-processing method focuses on managing the prediction results. This is due to the challenges and problems faced by the fine-tuned LM's generated outcomes. We carefully analyze the outcome and develop the following rules to ensure the model's outcome meets the requirements of the PPS-EMR competition.

Prediction Results do not Appear in the Input Text
When encountering situations where the predicted results do not appear in the input tokenized text, a comparison is further made with the original text to verify the relevance of the predicted results. If the extracted result remains unmatched, the result is removed (Fig. 2).

433475.RDC   MEDICALRECORD: 432175.RAC

Fig. 2.   Prediction results not appearing in the input text.

Duplicate Output of Prediction Results
The outputs of the fine-tuned LM are added into a set, which is a data structure that automatically eliminates duplicate elements, to ensure the uniqueness of recognized items (Fig. 3).

433475.RDC   MEDICALRECORD: 433475.RDC MEDICALRECORD: 433475.RDC MEDICALRECORD: 433475.RDC

Fig. 3.   Duplicate output of prediction results

Rule-Based Temporal Information Normalization

We notice that the fine-tuned model is occasionally "lazy" to provide the temporal normalization information. To avoid occasional errors, rule-based normalization methods are developed to enhance the recall of date-related information normalization (Fig. 4).

Last edited : 25/3/2064 Page: 2   DATE: 25/3/2064

Fig. 4. Rule-based temporal information normalization

Pattern-based Age-Related SHI Recognition.

To increase the recall of age-related SHIs, we developed a pattern-based method to recognize numerical values mentioned in the input text that match pre-defined patterns. The additional matched age-related SHIs are merged with the original prediction results.

Development Environment

To maintain consistency in training results throughout the entire training process, the system and environment as shown in Table 1 utilized Windows 11 2H22 version and Python 3.8.10.

The Python code utilized for this project references various packages and their versions as shown in Table 1. The following will provide a detailed explanation of the role of each package in this training.

Table 1. Package and Environment

Package	Version
Torch	2.0.1 + cu118
Transformer	4.35.0
Tqdm	4.66.1
Datasets	2.14.6
Random Re	

3 Results and Conclusion

In the first experiment shown in Fig. 5, we focused on evaluating and comparing the performance of three different scale models: Pythia-70m-deduped, Pythia-410m-deduped, and Pythia-1B-deduped to explore how the size of the model impacts its performance. Evaluations were conducted on the validation set released by PPS-EMR under the same preprocessing and post-processing configurations.

Coding Type	Precision	Recall	F-measure	Support
IDNUM	0.9522184	0.9481733	0.9501916	1177
MEDICALRECORD	0.9786477	0.9769094	0.9777778	563
PATIENT	0.7504244	0.8110092	0.7795414	545
AGE	1	0.9122807	0.9541284	57
CITY	0.9696048	0.9465876	0.9579579	337
STATE	0.9920319	0.8032258	0.8877006	310
STREET	0.8405797	0.5420561	0.6590909	321
ZIP	0.989547	0.8738462	0.9281046	325
DEPARTMENT	0.9732143	0.8934426	0.9316239	366
HOSPITAL	0.8056042	0.7757167	0.790378	593
DOCTOR	0.9238393	0.8083067	0.86222	2191
ORGANIZATION	0.4117647	0.2916667	0.3414634	24
PHONE	1	1	1	2
COUNTRY	0	0	0	2
LOCATION-OTHER	1	1	1	4
DATE	0.9826478	0.9461634	0.9640605	1616
TIME	0.9497042	0.7679426	0.8492064	418
DURATION	1	0.5	0.6666667	6
Micro-avg. F	0.9288417	0.8592074	0.8926686	8857
Macro-avg. F	0.8622128	0.7665182	0.8115543	8857

Temporal Type	Precision	Recall	F-measure	Support
DATE	0.8306082	0.7858911	0.8076311	1616
TIME	0.9345794	0.7177033	0.811908	418
DURATION	0.6666667	0.3333333	0.4444444	6
Micro-avg.	0.848354	0.7705882	0.8076034	2040
Macro-avg.	0.8106182	0.6123093	0.6976449	2040

(a)

Coding Type	Precision	Recall	F-measure	Support
IDNUM	0.9540817	0.953271	0.9536762	1177
MEDICALRECORD	0.9874101	0.9751332	0.9812332	563
PATIENT	0.8466523	0.7192661	0.7777778	545
AGE	1	0.8947368	0.9444444	57
CITY	0.9784616	0.9436202	0.9607251	337
STATE	1	0.8064516	0.8928571	310
STREET	0.7985612	0.6915888	0.7412354	321
ZIP	0.9964285	0.8584616	0.922314	325
DEPARTMENT	0.9736071	0.9071038	0.9391797	366
HOSPITAL	0.8170732	0.7908937	0.8037704	593
DOCTOR	0.9042386	0.788681	0.8425158	2191
ORGANIZATION	0.368421	0.2916667	0.3255814	24
PHONE	1	1	1	2
COUNTRY	0	0	0	2
LOCATION-OTHER	1	1	1	4
DATE	0.9947644	0.9405941	0.9669211	1616
TIME	0.9634146	0.7559808	0.847185	418
DURATION	1	0.1666667	0.2857143	6
Micro-avg. F	0.935081	0.853788	0.8925874	8857
Macro-avg. F	0.8657285	0.7491176	0.8032127	8857

Temporal Type	Precision	Recall	F-measure	Support
DATE	0.8335527	0.7840347	0.8080357	1616
TIME	0.9525316	0.7200957	0.8201635	418
DURATION	0	0	0	6
Micro-avg.	0.8535656	0.7686275	0.8088728	2040
Macro-avg.	0.5953614	0.5013768	0.5443421	2040

(b)

Fig. 5. (a) Validation results of Pythia-70m-deduped, (b) Validation results of Pythia-410m-deduped, (c) Validation results of Pythia-1B-deduped

Coding Type	Precision	Recall	F-measure	Support
IDNUM	0.9430009	0.9558199	0.9493671	1177
MEDICALRECORD	0.9873418	0.9698046	0.9784946	563
PATIENT	0.8262806	0.6807339	0.7464789	545
AGE	1	0.8421053	0.9142857	57
CITY	0.9906833	0.9465876	0.9681336	337
STATE	0.9959677	0.7967742	0.8853047	310
STREET	0.800738	0.6760125	0.7331081	321
ZIP	0.9929578	0.8676923	0.9261084	325
DEPARTMENT	0.9821959	0.9043716	0.9416785	366
HOSPITAL	0.8114187	0.7908937	0.8010248	593
DOCTOR	0.9192246	0.7790963	0.8433794	2191
ORGANIZATION	0.6470588	0.4583333	0.5365854	24
PHONE	1	1	1	2
COUNTRY	0	0	0	2
LOCATION-OTHER	1	1	1	4
DATE	0.9915803	0.947401	0.9689873	1616
TIME	0.9573171	0.7511961	0.841823	418
DURATION	1	0.5	0.6666667	6
Micro-avg. F	0.936427	0.8498363	0.8910329	8857
Macro-avg. F	0.8803203	0.7703789	0.8216884	8857

Temporal Type	Precision	Recall	F-measure	Support
DATE	0.8334422	0.7896039	0.810931	1616
TIME	0.9458598	0.7105263	0.8114754	418
DURATION	1	0.5	0.6666667	6
Micro-avg.	0.8528138	0.772549	0.8106996	2040
Macro-avg.	0.926434	0.6667101	0.7754012	2040

(c)

Fig. 5. (*continued*)

The average F-score comparison of the developed models on the validation set for the first subtask is shown in Fig. 7. We can observe that the performance gap between Pythia-1B-deduped and Pythia-70m-deduped in the first subtask "SHI Extraction" was minimal. This suggests that even with an increase in model parameters from 70m to 1B, the performance improvement was not significant. It implies that merely increasing the model size might not be sufficient to yield the expected benefit. However, for the results of the "temporal information normalization" subtask depicted in Fig. 8, we can observe a significant improvement in the Pythia-1B-deduped model, especially in the DURATION category, indicating a notable improvement in the normalization task (Fig. 6).

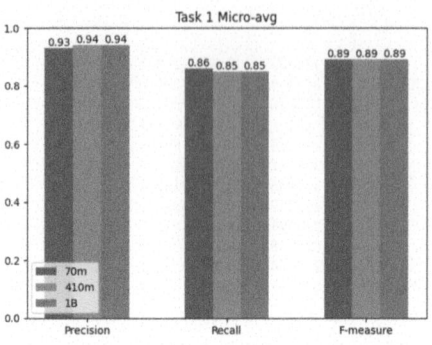

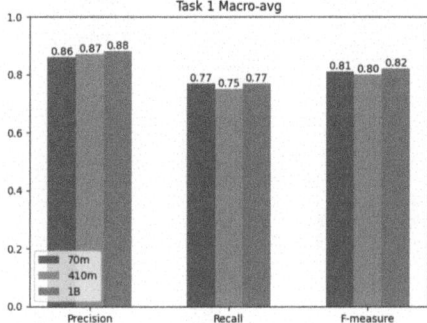

Fig. 6. The micro-avg. and macro-avg. F-measure comparisons between different-size models on the validation set for the first subtask.

Subsequently, we utilized the Pythia-70m-deduped, Pythia-410m-deduped, and Pythia-1B-deduped models in this competition, while also checking whether the prediction results matched the training validation results. The outcomes of the competition tests can be seen in Fig. 8.

According to the test set results, it is observed that the performance differences are relatively minimal in comparison to the validation set (Figs. 9 and 10).

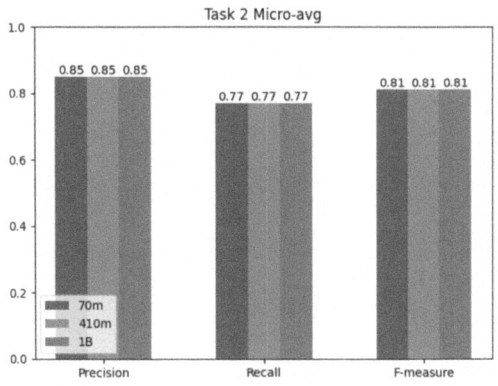

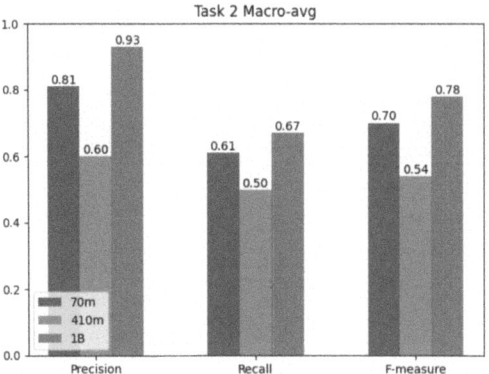

Fig. 7. The micro- and macro-avg. F-measure comparison between different model sizes on the validation set for the second task.

Our preliminary experiment results suggest that even an increase in the number of parameters does not necessarily lead to a significant improvement in the performance of the SHI recognition. Considering practical applications, if necessary, the Pythia-70m-deduped model might be a more pragmatic choice. Compared to Pythia-1B-deduped, it significantly reduces training and prediction time while offering comparably close performance.

Coding Type	Precision	Recall	F-measure
MEDICALRECORD	0.7878465	0.9892905	0.8771513
PATIENT	0.7723036	0.8100559	0.7907293
IDNUM	0.878125	0.9278302	0.9022936
DATE	0.9237829	0.9414396	0.9325277
DOCTOR	0.963806	0.7763751	0.8599967
CITY	0.97151	0.9142091	0.941989
STREET	0.9117647	0.5406977	0.6788321
ZIP	0.9869281	0.8555241	0.9165402
DEPARTMENT	0.8882353	0.7207637	0.7957839
HOSPITAL	0.7875417	0.590985	0.6752504
STATE	0.9961538	0.7801205	0.875
ORGANIZATION	0.4920635	0.4189189	0.4525548
DURATION	1	0.6666667	0.8
TIME	0.8192771	0.4340425	0.5674548
AGE	0.974359	0.7450981	0.8444445
LOCATION-OTHER	0.5	0.1666667	0.25
PHONE	1	1	1
SET	0	0	0
Micro-avg. F	0.8925403	0.812255	0.8505072
Macro-avg. F	0.8140943	0.6821492	0.742304

Temporal Type	Precision	Recall	F-measure
DATE	0.8267819	0.7783652	0.8018433
TIME	0.7254902	0.3148936	0.4391691
DURATION	0.625	0.4166667	0.5
SET	0	0	0
Micro-avg.	0.817966	0.7016293	0.7553444
Macro-avg.	0.544318	0.3774814	0.4458018

(a)

Coding Type	Precision	Recall	F-measure
MEDICALRECORD	0.7838126	0.9852744	0.8730723
PATIENT	0.8625429	0.7011173	0.7734976
IDNUM	0.8817345	0.9495283	0.9143766
DATE	0.8967413	0.9288329	0.912505
DOCTOR	0.9644167	0.7902014	0.8686601
CITY	0.979885	0.9142091	0.9459084
STREET	0.9182156	0.7180232	0.8058727
ZIP	0.9966443	0.8413598	0.9124424
DEPARTMENT	0.8948864	0.75179	0.8171207
HOSPITAL	0.7970688	0.5901502	0.6781775
STATE	0.996139	0.7771084	0.8730964
ORGANIZATION	0.4029851	0.3648649	0.3829787
DURATION	1	0.4166667	0.5882353
TIME	0.8205128	0.2723404	0.4089457
AGE	1	0.8431373	0.9148936
LOCATION-OTHER	0.5	0.1666667	0.25
PHONE	1	1	1
SET	0	0	0
Micro-avg. F	0.8954437	0.8098716	0.8505107
Macro-avg. F	0.8164213	0.6672929	0.7343627

Temporal Type	Precision	Recall	F-measure
DATE	0.8261821	0.7673851	0.7956989
TIME	0.7734375	0.2106383	0.3311037
DURATION	0.8	0.3333333	0.4705883
SET	0	0	0
Micro-avg.	0.8233347	0.6754922	0.7421219
Macro-avg.	0.5999049	0.3278392	0.4239797

(b)

Fig. 8. (a) Test set results of Pythia-70m-deduped, (b) Test set results of Pythia-410m-deduped, (c) Test set results of Pythia-1B-deduped.

Coding Type	Precision	Recall	F-measure	Support
MEDICALRECORD	0.7867804	0.9879518	0.8759644	747
PATIENT	0.8402662	0.7053072	0.7668945	716
IDNUM	0.8851233	0.9485849	0.9157559	2120
DATE	0.9058454	0.9389996	0.9221245	2459
DOCTOR	0.9589345	0.7790803	0.8597015	3327
CITY	0.9853373	0.9008043	0.9411765	373
STREET	0.9557195	0.752907	0.8422764	344
ZIP	0.9933993	0.8526912	0.9176829	353
DEPARTMENT	0.9266862	0.7541766	0.8315789	419
HOSPITAL	0.7945055	0.6035058	0.6859583	1198
STATE	0.9961686	0.7831326	0.8768971	332
ORGANIZATION	0.4929577	0.472973	0.4827586	74
DURATION	1	0.5	0.6666667	12
TIME	0.9012346	0.3106383	0.4620253	470
AGE	0.974359	0.7450981	0.8444445	51
LOCATION-OTHER	0.5	0.1666667	0.25	6
PHONE	1	1	1	1
SET	0	0	0	5
Micro-avg. F	0.8985643	0.8131775	0.8537412	13007
Macro-avg. F	0.8276287	0.6779177	0.7453296	13007

Temporal Type	Precision	Recall	F-measure	Support
DATE	0.8293633	0.7787719	0.8032718	2459
TIME	0.7191781	0.2234043	0.3409091	470
DURATION	1	0.5	0.6666667	12
SET	0	0	0	5
Micro-avg.	0.8232426	0.6877121	0.7493989	2946
Macro-avg.	0.6371354	0.375544	0.4725531	2946

Fig. 8. (*continued*)

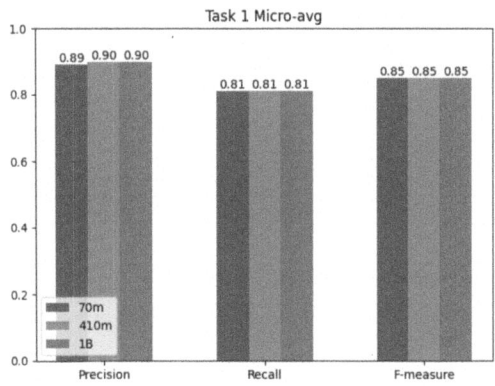

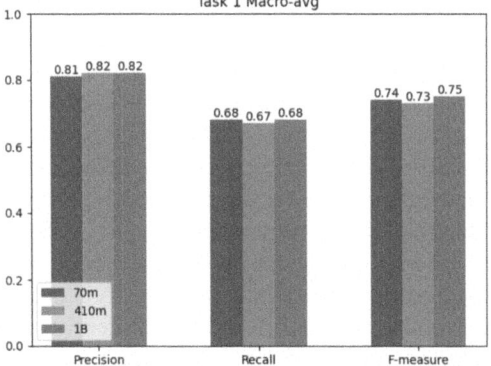

Fig. 9. Performance comparison on the test set for the first subtask

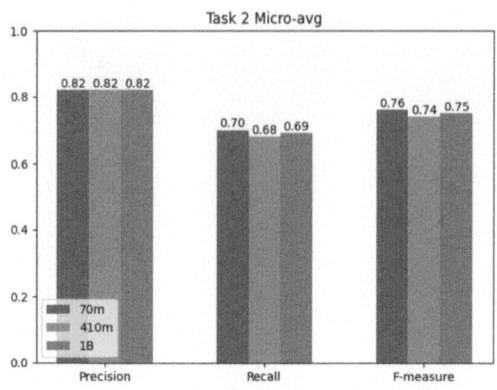

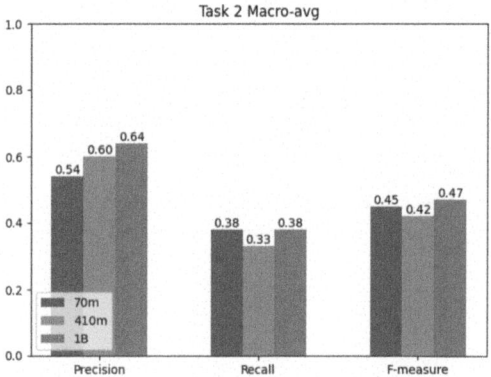

Fig. 10. Performance comparison on the test set for the second subtask.

4 Discussion

In this study, we conducted a comprehensive evaluation of the Pythia model's application in the de-identification and normalization of electronic medical records, aiming to protect patient privacy while enhancing the efficiency of medical data management automation. Our results indicate that merely increasing the model size does not significantly improve the effectiveness of identifying protected health information from Electronic Medical Record (EMR). However, for the task of temporal information normalization, fine-tuned large language models showed significant performance improvements. This study suggests that fine-tuning and optimization for specific tasks are more crucial than simply increasing model parameters. We also explored the model's efficacy and outcomes in handling sensitive medical data, finding that choosing a model of appropriate size may be more practical for real-world applications, significantly reducing training and prediction time while maintaining similar performance. Moreover, we discussed detailed post-processing methods of model predictions, crucial for meeting specific competition requirements and enhancing model practicality. Through this research, we aim to provide valuable insights and methodologies for the field of de-identification and normalization research in electronic medical records.

Acknowledgments. We express our gratitude to Professor Hong-Jie Dai and the organizers of the AI CUP 2023 Privacy Protection and Standardization of Electronic Medical Record Competition for their invaluable guidance and feedback on our work. Additionally, we recognize the IW-DMRN workshop (https://www.sredhconsortium.org/sredh-competitions/sredhai-cup-2023/ 2024-iw-dmrn) and the SREDH Consortium (https://www.sredhconsortium.org/) for their valuable contributions to our research endeavors.

Disclosure of Interests. The authors have no competing interests.

References

1. Wu, X., Duan, R., Ni, J.: Unveiling security, privacy, and ethical concerns of ChatGPT. J. Inf. Intell. **2**(2), 102–115 (2023)
2. Hu, D., Liu, B., Zhu, X., Lu, X., Wu, N.: Zero-shot information extraction from radiological reports using ChatGPT. Int. J. Med. Inf. **183**, 105321 (2024)
3. Wu, T., et al.: A brief overview of ChatGPT: the history, status quo, and potential future development. IEEE/CAA J. Automatica Sinica **10**(5), 1122–1136 (2023)
4. Qu, Y., Liu, P., Song, W., Liu, L., Cheng, M.: A text generation and prediction system: pre-training on new corpora using BERT and GPT-2. In: 2020 IEEE 10th International Conference on Electronics Information and Emergency Communication (ICEIEC), pp. 323–6. IEEE (2020)
5. Liu, J., Gupta, S., Chen, A., Wang, C.-K., Mishra, P., Dai, H.-J., et al.: OpenDeID pipeline for unstructured electronic health record text notes based on rules and transformers: de-identification algorithm development and validation study. J. Med. Internet Res. **25**, e48145 (2023)
6. Chen, A., Jonnagaddala, J., Nekkantti, C., Liaw, S.-T.: Generation of surrogates for de-identification of electronic health records. MEDINFO 2019: health and Wellbeing e-Networks for All. IOS Press, pp. 70–73 (2019)
7. Ilić D. Unveiling the general intelligence factor in language models: a psychometric approach (2023). https://arxiv.org/abs/2310.11616
8. SREDH Consortium. https://www.sredhconsortium.org/sredh-consortium
9. SREDH Consortium. SREDHAI Cup 2023/2024 IW-DMRN. https://www.sredhconsortium. org/sredh-competitions/sredhai-cup-2023/2024-iw-dmrn
10. Mir, T.H., Yang, H.P., Chou, Y.Y., Teng, Y.C., Liao, W.H., Lin, Y.C., et al. De-identification and temporal normalization of electronic health record notes using large language
11. Models: The SREDH/AI-Cup 2023 De-identification Competition. In: 2024 International Workshop on De-identification of Electronic Medical Record Notes. (2024). Kaohsiung, Taiwan: Springer Nature. https://www.sredhconsortium.org/sredh-competitions/sredhai-cup-2023
12. Jonnagaddala, J., Chen, A., Batongbacal, S., Nekkantti, C.: The OpenDeID corpus for patient de-identification. Sci. Rep. **11**(1), 19973 (2021)
13. Alla, N.L.V., et al.: Cohort selection for construction of a clinical natural language processing corpus. Comput. Methods Programs Biomed. Update **1**, 100024 (2021)
14. ISLAB. Privacy protection and medical data standardization competition: decoding clinical cases and letting data tell stories (2023). https://codalab.lisn.upsaclay.fr/competitions/15425

A Two-Stage Fine-Tuning Procedure to Improve the Performance of Language Models in Sensitive Health Information Recognition and Normalization Tasks

Pin-Sen Chiu[1]([⊠]) [ID], Bo-Wei Hou[2] [ID], Yen-Ting Chen[3] [ID], and Shi-Hao Huang[4] [ID]

[1] Applied Intelligence Lab, Department of Electrical Engineering, National Kaohsiung University of Science and Technology, Kaohsiung 807618, Taiwan
`f113154137@nkust.edu.tw`

[2] Intelligent Control and Computer Vision Lab, Department of Electrical Engineering, National Kaohsiung University of Science and Technology, Kaohsiung 807618, Taiwan
`f113154124@nkust.edu.tw`

[3] Intelligent Automation Laboratory, Department of Electrical Engineering, National Kaohsiung University of Science and Technology, Kaohsiung 807618, Taiwan
`f113154125@nkust.edu.tw`

[4] Broadband Mobile Wireless Lab, Department of Electrical Engineering, National Kaohsiung University of Science and Technology, Kaohsiung 807618, Taiwan
`f113154132@nkust.edu.tw`

Abstract. In recent years, artificial intelligence (AI) technology has experienced flourishing development, and large-scale language models (LMs) are increasingly recognized as a pivotal direction for the future of AI in the field of intelligent healthcare. In the realm of clinical medical applications, there is an increasing focus on eliminating patient privacy information from textual electronic health records within medical institutions. To address this, we have developed a two-stage fine-tuning procedure which employs the Pythia model for training purposes, enabling the model to effectively extract and normalize patient-related privacy information. Furthermore, ChatGPT is utilized for data augmentation to generate additional training data, enhancing the overall robustness of the system.

Keywords: Natural Language Processing · Sensitive Health Information · Pythia · ChatGPT · Language Model

1 Introduction

In recent years, the rapid advancement of artificial intelligence (AI) technology has seen the integration of large-scale language models (LMs) into various applications. Industry giants like OpenAI, Microsoft, and Google have prominently featured models such as ChatGPT, Bing Search, and Bard, showcasing their potential across different domains. Particularly in clinical healthcare, these models play a pivotal role, suggesting a significant direction for AI-driven smart healthcare in the future. However, concerns about user privacy have surfaced due to interactions with such models, potentially leading to the inadvertent disclosure of sensitive information.

J. Jonnagaddala et al. (Eds.): IW-DMRN 2024, CCIS 2148, pp. 202–212, 2025.
https://doi.org/10.1007/978-981-97-7966-6_15

Our participation in the AI-CUP 2023 competition [1, 6, 7] has offered valuable insights into these challenges and provided methodologies for addressing them. To effectively utilize electronic health records (EHRs) within healthcare institutions for medical research and smart healthcare applications, it is essential to remove patient privacy information from the texts. Given the various privacy issues stemming from the use of EHRs and large-scale language models, this competition challenges participants' artificial intelligence development skills by providing de-identified EHR datasets from real healthcare institutions. Participants are assigned two sub-competition missions, which include Patient Privacy Information Extraction and Temporal Data Normalization.

Furthermore, as healthcare institutions worldwide transition towards digitalization, EHRs have become indispensable sources for clinical data analysis. However, these textual EHRs often contain sensitive patient information. Details like a patient's date of birth, appointment times, and attending physicians are frequently intertwined, posing a risk of inferring the real identities of patients and compromising their privacy. Current clinical Natural Language Processing (NLP) in Electronic Health Records (EHRs) has been widely studied [9]. Recent advancements have incorporated transformer-based models to improve de-identification performance [10], aiming to address challenges such as surrogate generation and document context preservation [8].

To tackle these challenges, this paper proposes a two-stage fine-tuning procedure using the Pythia model [11] as the primary framework for training and prediction. This approach aims to effectively extract patient privacy information while normalizing temporal data within textual EHRs. Additionally, leveraging ChatGPT, developed by OpenAI, for data augmentation enhances the system's robustness through the generation of supplementary training data.

In our research, we referred to the method introduced in [2] and further adopted a similar low-rank adaptation (LoRA) approach. Specifically, we employed the technique of Fine-tuning followed by LoRA to enhance the performance of large language models in sensitive health information recognition and normalization tasks.

In this approach, we first performed Fine-tuning on a pre-trained language model and then applied the low-rank weight update method of LoRA to better adapt to these specific tasks. The core concept of this method is to apply low-rank weight updates to the model after Fine-tuning to further improve its performance while maintaining computational efficiency.

Preliminary experimental results demonstrate that this method shows superior performance in sensitive health information recognition and normalization tasks compared to using Fine-tuning or LoRA alone. Overall, this study explores the combination of Fine-tuning and LoRA methods to enhance the performance of large language models in sensitive health information recognition and normalization tasks.

2 Related Work

In the paper "Data augmentation using pre-trained transformer models" [3] explores the effectiveness of various transformer-based pre-trained models, such as GPT-2, BERT, and BART, for conditional data augmentation in natural language processing (NLP) tasks. The authors propose a unified approach to use different pre-trained transformer models for data augmentation across sentiment classification, intent classification, and question classification. The study demonstrates that prepending class labels to text sequences is a simple yet effective way to condition pre-trained models for data augmentation. In a low-resource setting with only 10 training examples per class, pre-trained seq2seq models outperform other data augmentation methods, showcasing their significance in improving classification performance. The paper contributes by implementing a seq2seq pre-trained model for data augmentation, experimentally comparing various methods, and providing practical guidelines for utilizing different pre-trained models.

The paper "Lora: Low-rank adaptation of large language models" [4] introduces a novel approach called Low-Rank Adaptation (LoRA) for efficiently adapting pre-trained language models to various downstream tasks. Traditional fine-tuning involves updating all parameters, leading to increased model size, especially in large-scale models like GPT-3. LoRA proposes a low-rank decomposition method, where only a small number of task-specific parameters are trained, significantly reducing storage and computational requirements. This approach allows for efficient task switching, lowers the hardware entry barrier, and introduces no inference latency compared to fully fine-tuned models. LoRA is shown to be compatible with various prior methods, making it a versatile solution for adapting pre-trained models.

3 Methodology

3.1 Dataset Pre-processing

Sensitive health information is frequently obscured form unstructured electronic health records to safeguard patient privacy and confidentiality. The OpenDeID v2 corpus comprises of 3,244 reports out of which 2100 reports are from OpenDeID v1 corpus [5] aims to support the advancement of automated techniques for obscuring such sensitive information from unstructured electronic health records. Using both the first stage and second-stage datasets from AICUP-2023 competition [1], the main steps include reading each line in every document and calculating its starting and ending positions. The generated data consists of five main fields: "file" (file name), "contents start" (start position of content), "contents end" (end position of content), "contents" (content), and "labels" (extracted data). The "labels" field covers data for all categories and SHI (Personally Identifiable Information), with "PHI: NULL" indicating no extracted data (Fig. 1).

In the dataset, due to the fewer data entries for the categories COUNTRY, LOCATION-OTHER, PHONE, and DURATION, special handling is performed. In particular, for COUNTRY and LOCATION-OTHER categories, the corresponding data rows are directly removed to minimize their impact on the overall model. For PHONE and DURATION categories, we utilized the ChatGPT model by applying it to reference existing data and generate new text in a similar format to increase diversity and coverage. This approach effectively increases the amount of data for PHONE and DURATION categories, thereby improving the model's performance in these categories.

Additionally, for extracting data in the prediction process, regular expressions and dictionaries are utilized for COUNTRY and LOCATION-OTHER categories to efficiently capture the required information. This approach is chosen because DURATION

	A file	B contents_start	C contents_end	D contents	E labels
1	file	contents_start	contents_end	contents	labels
2	10	0	1		PHI: NULL
3	10	1	25	Episode No: 09F016547J	IDNUM 14 24 09F016547J
4	10	25	36	091016.NMT	MEDICALRECORD 25 35 091016.NMT
5	10	36	37		PHI: NULL
6	10	37	52	SIZAR, HOWARD	PATIENT 37 50 SIZAR, HOWARD
7	10	52	70	Lab No: 09F01654	IDNUM 61 69 09F01654
8	10	70	78	Runford	STREET 70 77 Runford
9	10	78	97	RENMARK TAS 5084	CITY 78 85 RENMARK / STATE 87 90 TAS / ZIP 92 96 5084
10	10	97	114	Specimen: Tissue	PHI: NULL
11	10	114	132	D.O.B: 24/8/1993	DATE 122 131 24/8/1993 1993-08-24
12	10	132	140	Sex: M	PHI: NULL
13	10	140	171	Collected: 28/08/2013 at 08:2	TIME 151 170 28/08/2013 at 08:26 2013-08-28T08:26
14	10	171	230	Location: St Vincent-BATLO	DEPARTMENT 182 192 St Vincent / HOSPITAL 193 229 BATLOW/ADELONG MULTI PURPOSE SERVICE
15	10	230	250	DR JAXON AL-KARSTEN	DOCTOR 233 249 JAXON AL-KARSTEN

Fig. 1. Preprocessed training data

content lacks a specific format, PHONE training performs well without specific regular expression methods, and COUNTRY and LOCATION-OTHER are handled oppositely.

In summary, this comprehensive approach, combining first-stage and second-stage datasets, removing specific category rows, and augmenting other categories through ChatGPT, enables better data handling, enhancing the model's performance and accuracy. This data processing workflow provides more complete and effective resources for subsequent model training and predictions.

3.2 Model and Fine-Tuning Setting

The pre-trained model used was the pythia-160m model, a language generation model based on the Transformers framework, adopting the GPT-NeoX architecture. It is a self-regressive language model with each layer consisting of self-attention mechanisms and multi-layer perceptrons, enabling the model to handle long-range semantic dependencies and perform well in natural language processing tasks.

The model's underlying structure includes an Embedding layer, multiple GPT-NeoX layers, and a final Linear layer. The Embedding layer maps inputs (tokens) to a 768-dimensional vector space, while the GPT-NeoX layers capture semantic features. The final Linear layer maps the model's output back to the size of the model's vocabulary, completing the entire generation process. The model benefits from a large amount of data during the learning process and can generate fluent and logical natural language text.

In terms of model training, we defined the tokens in Table 1 and adopted the following training format:

bos_token sentence_content sep_token gold annotations eos_token.

We adjusted the number of epochs by observing the average training loss to achieve better results. After observing the results, we found that the model's performance is more stable and effective when the loss value is maintained between 0.5 and 0.6.

Table 1. Token definition.

Token Type	Token Value
bos_token	< lendoftextl>
sep_token	\n\n####\n\n
eos_token	< lendl>
pad_token	< lpadl>

In terms of model training, we conducted two-steps fine-tuning procedure of the model. We first use each data entry with contents and gold labels for training over 13 epochs. Subsequently, we employed the LoRA (Low-Rank Adaptation of Large Language Models) method for parameter-efficient fine-tuning, which involves freezing the weights of the original pre-trained model and introducing a small LoRA network in specific layers for training. We performed 5 additional epochs of LoRA training on our model with the highest F-scores after fine-tuning and selected the number of epochs with the best performance. We observed that the hybrid use of fine-tuning and LoRA strategies ensures the model adapts well to the data during the training process. This training strategy provides effective support for optimizing the model (Table 2).

Table 2. Example of training data

<lendoftextl>Please contact our research team at (1111 2222) for collaboration inquiries.
####
PHONE 1111 2222
<lEND|>
<lendoftextl>Right brachial plexus tumour. ?Neurofibroma. ?Schwannoma.
####
SHI: NULL<lEND|>
<lendoftextl>(TO: RF;BL/ec 29.10.63)
####
DOCTOR RF
DOCTOR BL
DATE   29.10.63 2063-10-29
<lEND|>

Finally, the choice of optimizer is crucial for achieving good model performance. We used the AdamW optimizer, an improvement over the Adam optimizer, introducing weight decay to prevent model overfitting and enhance training stability.

4 Results

4.1 Performance on the Development Set

We fine-tuned the Pythia160md model using preprocessed data and tested it on validation data. Through Fig. 2, it was observed that the score for task one reached its highest point within the first 2 epochs among the epochs 1 to 13. Notably, within the initial 2 epochs, the model attains its highest score for task one. This underscores the Pythia-160m model's rapid adaptation to the specific demands of extracting patient privacy information from healthcare records.

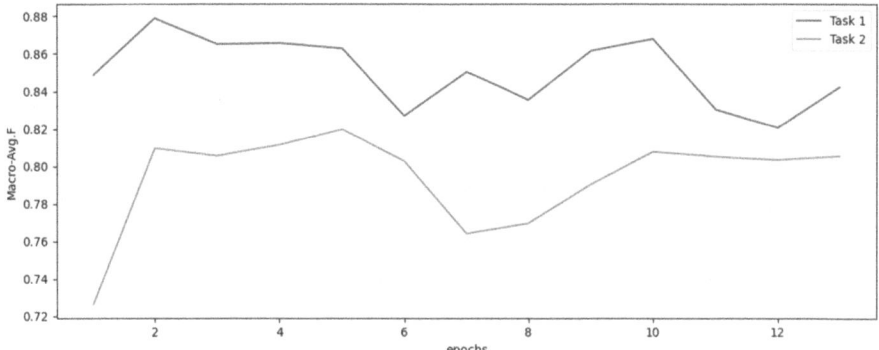

Fig. 2. Scores of Fine-tuning

After fine-tuning for 2 epochs, training with LoRA was found to further enhance the scores. The fine-tuning process, as depicted in Fig. 3, showcases the effectiveness of our approach. The subsequent LoRA fine-tuning demonstrates a noticeable improvement in scores. This iterative refinement process underscores the model's adaptability and the positive influence of incorporating additional training techniques. Importantly, LoRA selection is based not solely on achieving the highest score, but rather on choosing the most significant score improvement. Consequently, in our experimentation, we opted for a strategic combination of fine-tuning for 2 epochs followed by LoRA for 1 epoch, recognizing the synergy between these phases for optimal performance.

Figure 4 shows the effectiveness of applying ChatGPT for data augmentation, especially in addressing SHI categories with limited training instances. We applied the augmentation method to increase the amount of data for specific categories such as PHONE and DURATION, contributing to the improved model performance for these SHI types.

Under the conditions of fine-tuning for 2 epochs followed by LoRA for 1 epoch, it was found, as depicted in Fig. 4, that augmenting the training data for the PHONE and DURATION categories in both the original dataset and the increased dataset results in a significant improvement in scores.

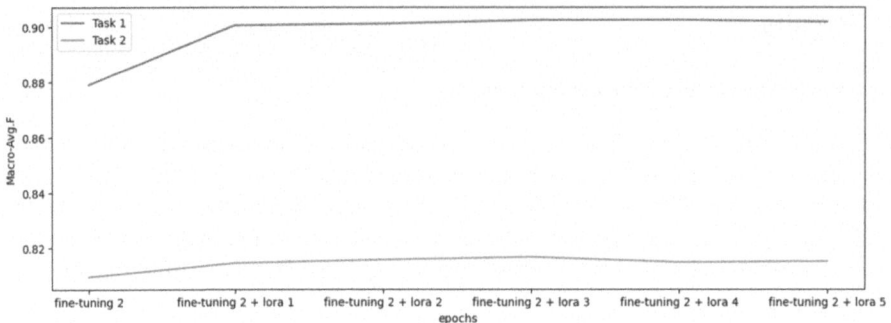

Fig. 3. Macro-avg. F-scores comparison of the proposed two-stage fine-tuning procedure.

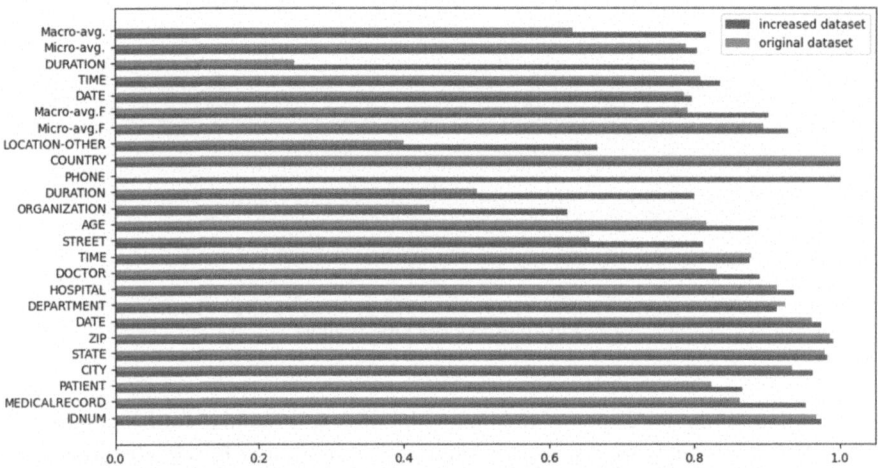

Fig. 4. Performance comparison of the models fine-tuned with the original and augmented datasets.

4.2 Performance on the Test Set

In Fig. 5, we present the scores of our trained model on the test set. The SET category, which was not present in the training data, shows performance consistent with its absence in the training set. However, a noticeable regression is observed in the TIME category, where the model's performance has declined significantly compared to its originally good performance. This could be attributed to differences between the training data and the test set, suggesting a potential need for data augmentation for the SHI types. Addressing this issue is an area for improvement in our future work.

```
| Coding Type    | Precision  | Recall    | F-measure   | Support  |
|--------------- |----------- |:--------- |------------ |--------- |
| MEDICALRECORD  | 0.7921478  | 0.91834   | 0.850589    | 747      |
| PATIENT        | 0.8691099  | 0.9273743 | 0.8972973   | 716      |
| IDNUM          | 0.9746169  | 0.9599057 | 0.9672053   | 2120     |
| DATE           | 0.9410329  | 0.9410329 | 0.9410329   | 2459     |
| DOCTOR         | 0.9644154  | 0.7983168 | 0.8735405   | 3327     |
| STREET         | 0.8925081  | 0.7965117 | 0.8417819   | 344      |
| CITY           | 0.9719101  | 0.9276139 | 0.9492455   | 373      |
| STATE          | 0.9938272  | 0.9698795 | 0.9817073   | 332      |
| ZIP            | 0.991404   | 0.98017   | 0.985755    | 353      |
| DEPARTMENT     | 0.8004695  | 0.8138425 | 0.8071006   | 419      |
| HOSPITAL       | 0.8892889  | 0.8247079 | 0.8557817   | 1198     |
| DURATION       | 0.9        | 0.75      | 0.8181818   | 12       |
| TIME           | 0.4797508  | 0.3276596 | 0.3893805   | 470      |
| AGE            | 0.8974359  | 0.6862745 | 0.7777778   | 51       |
| ORGANIZATION   | 0.4303797  | 0.4594595 | 0.4444444   | 74       |
| LOCATION-OTHER | 1          | 0.1666667 | 0.2857143   | 6        |
| SET            | 0          | 0         | 0           | 5        |
| PHONE          | 0          | 0         | 0           | 1        |
| Micro-avg. F   | 0.9143953  | 0.8614592 | 0.8871383   | 13007    |
| Macro-avg. F   | 0.7660165  | 0.6804309 | 0.7206917   | 13007    |
|--------------- |----------- |:--------- |------------ |--------- |
|--------------- |----------- |:--------- |------------ |--------- |
| Temporal Type  | Precision  | Recall    | F-measure   | Support  |
|--------------- |----------- |:--------- |------------ |--------- |
| DATE           | 0.8085566  | 0.7608784 | 0.7839933   | 2459     |
| DURATION       | 0.8888889  | 0.6666667 | 0.7619048   | 12       |
| TIME           | 0.3246753  | 0.106383  | 0.1602564   | 470      |
| SET            | 0          | 0         | 0           | 5        |
| Micro-avg.     | 0.7787646  | 0.6547862 | 0.7114143   | 2946     |
| Macro-avg.     | 0.5055302  | 0.383482  | 0.4361284   | 2946     |
|--------------- |----------- |:--------- |------------ |--------- |
```

Fig. 5. Performance of the developed model on the test set.

5   Discussion

Our study introduces a novel approach that combines Fine-tuning with LoRA for enhancing the performance of large language models in sensitive health information recognition and normalization tasks. The observed improvement in task-specific performance underscores the effectiveness of leveraging both techniques in adapting models to specialized domains like healthcare. By iteratively refining the model's weights through Fine-tuning and applying low-rank weight updates via LoRA, we achieve superior results without significantly increasing computational costs. This finding suggests that task-specific adaptation techniques are crucial for optimizing model performance in healthcare contexts.

However, it is essential to acknowledge the limitations and challenges associated with our approach. Firstly, our evaluation is confined to a specific set of tasks and datasets related to sensitive health information, which may limit the generalizability of our findings to other healthcare domains. Additionally, while our method shows promising results in controlled settings, its performance in real-world clinical environments may vary due to factors such as data heterogeneity and domain-specific nuances.

Despite these limitations, our study offers valuable insights into the practical implications of leveraging advanced NLP techniques in healthcare. Accurate recognition and normalization of sensitive health information from unstructured text data have significant implications for healthcare providers, researchers, and policymakers. Our findings suggest that by employing innovative NLP techniques like Fine-tuning with LoRA, we can enhance the efficiency and accuracy of processing health-related information, thereby facilitating tasks such as electronic health record management, clinical decision support, and patient communication.

Looking ahead, several avenues for future research and development emerge from our study. Firstly, further optimization and exploration of alternative adaptation techniques tailored specifically for healthcare-related tasks are warranted. Investigating different architectures, optimization algorithms, or incorporating domain-specific knowledge could yield additional performance improvements. Moreover, conducting more extensive evaluations on diverse healthcare datasets and collaborating with domain experts to ensure the practical relevance of proposed methods are crucial steps for advancing the field.

In conclusion, our study contributes to advancing knowledge in the field of NLP for healthcare by proposing an effective method for adapting large language models to sensitive health information tasks. Through comprehensive analysis and discussion, we provide insights into the implications of our findings and highlight avenues for future research and real-world applications. By addressing these challenges and leveraging innovative NLP techniques, we can continue to improve the quality and efficiency of healthcare services in the era of virtual care.

6 Conclusion

One of the primary challenges encountered in this study was the limited availability of training data for certain categories such as COUNTRY, LOCATION-OTHER, PHONE, and DURATION. This scarcity necessitated special handling, including data removal and augmentation, to ensure balanced training across all categories. Additionally, the variation in data format posed challenges, particularly in extracting DURATION content, which lacked a specific format. These challenges underscore the complexity of real-world data and the need for robust processing techniques.

The choice of model architecture and fine-tuning strategies significantly influenced the performance of the system. By employing the Pythia-160m model as the backbone and incorporating LoRA for parameter-efficient fine-tuning, we aimed to strike a balance between model adaptability and computational efficiency. However, determining the optimal number of epochs for fine-tuning and LoRA training proved to be non-trivial and required careful experimentation. Future studies could explore alternative model architectures and fine-tuning approaches to further improve performance.

Our study presumed that the combination of Pythia-160m and ChatGPT, along with meticulous data preprocessing, would result in an effective system for extracting patient privacy information. While our results support this presumption to a large extent, it is important to acknowledge that the performance of the system may vary depending on factors such as data quality and domain specificity. Moreover, the assumption that data augmentation using ChatGPT would sufficiently address the scarcity of training data for certain categories warrants further investigation.

Despite the challenges and presumptions, this study opens up several promising avenues for future research. One potential direction is the exploration of ensemble methods to combine predictions from multiple models, thereby enhancing overall performance and robustness. Additionally, incorporating domain-specific knowledge and ontologies into the model architecture could improve the accuracy of data extraction and normalization. Furthermore, collaboration with healthcare institutions to obtain annotated datasets and validate the system in real-world settings could provide valuable insights into its practical utility.

In conclusion, the integration of Pythia-160m and ChatGPT, coupled with a meticulous data processing strategy, results in a robust system for the extraction of patient privacy information from pathology reports. The model's adaptability, innovative augmentation techniques, and efficient data processing contribute to its overall effectiveness. This study not only showcases the potential of large-scale LMs in healthcare applications but also emphasizes the importance of thoughtful data handling and training approaches in ensuring model performance and reliability. As we navigate the intersection of AI and healthcare, the findings presented here pave the way for further advancements in intelligent healthcare solutions.

Acknowledgments. We extend our heartfelt gratitude to Professor Hong-Jie Dai for his invaluable guidance throughout the execution of our project. Additionally, we thank the organizing institutions of the AI CUP 2023 Privacy Protection and Standardization of Electronic Medical Record Competition: the Intelligent Systems Laboratory of the Department of Electrical Engineering at National Kaohsiung University of Science and Technology, the Department of Bioinformatics and Medical Engineering at Asia University, and the SREDH Association of Australia, for providing the platform and resources for this competition.

Disclosure of Interests. The authors have no competing interests.

References

1. AICup website - https://codalab.lisn.upsaclay.fr/competitions/15425
2. Xia, W., Qin, C., Hazan, E.: Chain of LoRA: efficient fine-tuning of language models via residual learning. arXiv preprint arXiv:2401.04151 (2024)
3. Kumar, V., Choudhary, A., Cho, E.: Data augmentation using pre-trained transformer models. arXiv preprint arXiv:2003.02245 (2020)
4. Hu, E.J., et al.: LoRA: Low-rank adaptation of large language models. arXiv preprint arXiv: 2106.09685 (2021)
5. Jonnagaddala, J., Chen, A., Batongbacal, S., Nekkantti, C.: The OpenDeID corpus for patient de-identification. Sci. Rep. **11**(1), 19973 (2021)
6. SREDH Website - https://www.sredhconsortium.org/
7. IW-DMRN website - https://www.sredhconsortium.org/sredh-competitions/sredhai-cup-2023/2024-iw-dmrn
8. Chen, A., Jonnagaddala, J., Nekkantti, C., Liaw, S.T.: Generation of surrogates for de-identification of electronic health records. In: MEDINFO 2019: Health and wellbeing e-networks for all, pp. 70–73. IOS Press (2019)
9. Alla, N.L.V., Chen, A., Batongbacal, S., Nekkantti, C., Dai, H.J., Jonnagaddala, J.: Cohort selection for construction of a clinical natural language processing corpus. Comput. Methods Programs Biomed. Update, **1**, 100024 (2021)

10. Liu, J., et al.: OpenDeID pipeline for unstructured electronic health record text notes based on rules and transformers: de-identification algorithm development and validation study. J. Med. Internet Res. **25**, e48145 (2023)
11. Biderman, S., et al.: Pythia: a suite for analyzing large language models across training and scaling. In: International Conference on Machine Learning, pp. 2397–2430. PMLR (2023)

Author Index

J. Jonnagaddala et al. (Eds.): IW-DMRN 2024, CCIS 2148, pp. 213–214, 2025.
https://doi.org/10.1007/978-981-97-7966-6

The manufacturer's authorised representative in the EU is Springer
Nature Customer Service Centre GmbH, Europaplatz 3, 69115 Heidelberg,
Germany. If you have any concerns regarding our products, please
contact ProductSafety@springernature.com

Printed and bound by CPI Group (UK) Ltd, Croydon, CR0 4YY
29/04/2026
02099537-0001